RENEWABLE ENERGY

Cleaner, fairer ways to power the planet

About the author

Danny Chivers is an environmental writer, researcher, professional carbon footprint analyst, activist and performance poet. He holds a BSc in Environmental Biology, an MSc in Nature, Science and Environmental Policy and an MProf in Leadership for Sustainable Development. He has carried out studies into the climate impact of electrical product manufacture, agriculture, food processing, international development NGOs, local authority carbon monitoring, retailers, offices and the UK government. He has created an interactive emissions model of the UK economy for *The Guardian* website, co-founded 'Cyc du Soleil' (Britain's first mobile solar-and-cycle-powered performance stage), taken part in the Climate Camps at Heathrow, Kingsnorth and central London, and reached the semi-finals of the BBC Radio 4 National Poetry Slam. He is the author of *The NoNonsense Guide to Climate Change: The Science, the Solutions, the Way Forward* (New Internationalist 2010).

About the New Internationalist

New Internationalist is an award-winning, independent media co-operative. Our aim is to inform, inspire and empower people to build a fairer, more sustainable planet.

We publish a global justice magazine and a range of books, both distributed worldwide. We have a vibrant online presence and run ethical online shops for our customers and other organizations.

- **Independent media:** we're free to tell it like it is – our only obligation is to our readers and the subjects we cover.
- **Fresh perspectives:** our in-depth reporting and analysis provide keen insights, alternative perspectives and positive solutions for today's critical global justice issues.
- **Global grassroots voices:** we actively seek out and work with grassroots writers, bloggers and activists across the globe, enabling unreported (and under-reported) stories to be heard.

NONONSENSE

RENEWABLE ENERGY

Cleaner, fairer ways to power the planet

Danny Chivers

New Internationalist

NONONSENSE

Renewable Energy

Cleaner, fairer ways to power the planet

Published in 2015 by
New Internationalist Publications Ltd
The Old Music Hall
106-108 Cowley Road
Oxford OX4 1JE, UK
newint.org

Reprinted in 2016

Cover design: Andrew Smith, asmithcompany.co.uk

Series editor: Chris Brazier
Series design by Juha Sorsa

Printed and bound in Great Britain by Bell & Bain Ltd, Glasgow
who hold environmental accreditation ISO 14001.

British Library Cataloguing-in-Publication Data.
A catalogue record for this book is available from the British Library.

Library of Congress Cataloging-in-Publication Data.
A catalog for this book is available from the Library of Congress.

ISBN 978-1-78026-243-7
(ISBN ebook 978-1-78026-244-4)

Contents

Foreword

Danny Chivers packs a lot in to 170 pages! From the types of renewable technologies on offer to the challenges of their implementation and suggested action steps for the reader, *NoNonsense Renewable Energy* offers a fantastic introduction to this essential topic.

The climate science could not be clearer: human influence on the climate system is substantial and is still growing. The 2014 report released by scientists from the IPCC (International Panel on Climate Change) said: 'If left unchecked, climate change will increase the likelihood of severe, pervasive and irreversible impacts for people and ecosystems.' The business-as-usual scenario is not an option if we are to avoid the most devastating effects of climate change; we need a rapid transition to a clean, efficient, renewables-based energy system.

Skillfully written to make the subject of renewable energies accessible and engaging, this book is a vital tool for anyone interested in climate-change solutions. From solar, wind, hydroelectric, heat pumps, wave and tidal to fuel crops and energy from waste, Danny Chivers covers the technologies on offer, their historical use and their potential in rolling out a 100-per-cent renewable future. Importantly, he also explores the costs, risks and drawbacks associated with each technology.

Along with the practical side of renewables, he also explores the potential for, and the barriers to, a clean-energy future. As the author points out, it is not technology that is standing in the way of making a transition to a sustainable society, but a political and economic system which, in its thirst for profit, overrides sustainable solutions.

'Who has power over the power?' is a key question asked by the book. Simply shifting power generation to a 100-per-cent renewable scenario replicates our current (unjust) energy system as long as large energy corporations remain in control of that power. The democratization of our energy and our politics needs to happen in parallel, and for that transition to happen there must be a decisive shift in power towards people, communities, small local businesses and workers.

The last chapter of the book, 'Making it happen', really is the icing on the cake; it is one thing to present useful information but it is truly wonderful when a book leaves you feeling empowered and ready to take action. With its suggestions for '5 ways forward' and a list of action groups to look up, the chapter provides plenty of tools to get started in making a change.

Society is standing at a crossroads, and change is coming whether we like it or not. We can either change by choice now, or change by force at some point in the future. Building a sustainable future is within our reach, but to do so we need to build a massive and informed movement for change. New Internationalist's *NoNonsense* guides have been game changers in setting out complex topics in an informative and accessible way. *NoNonsense Renewable Energy* is an essential read for anyone who is planning to be part of the energy revolution.

Kim Bryan
Centre for Alternative Technology, Machynlleth, Wales

Introduction: What is renewable energy and why do we need it?

> *'It's as if you live in a fantasy world where we can get all the energy we need from wind, solar, unicorn farts and rainbows.'*
>
> – online comment on an article criticizing the fossil-fuel industry

We're surrounded by more energy than we could ever use. Every hour, the sun pours more energy onto the Earth than the entire global economy uses in a year.[1] Every month, enough sunshine to power our society for centuries slips quietly through the atmosphere, reflecting off coal power station chimneys, lending a gleam to gas pipelines and giving sunburns to oil-rig workers.

The difficulty lies in capturing this solar bounty. For most of human history, our ancestors gathered their sunshine indirectly. For millions of years, the main route through which the sun's energy reached us was through the everyday miracle of photosynthesis (the chemical process in plants' leaves that uses the energy in sunlight to transform carbon dioxide and water into oxygen and sugars). As well as powering our own bodies via the food we (and other animals) eat, plants have been a source of burnable fuel for humanity and our ancestors for around two million years; wood is still the main source of heating and cooking energy for a third of the planet's people.[2]

Another major method for harnessing the power of the sun arrived around 6,000 years ago (4000 BCE), with the domestication of oxen, buffalo and llamas as beasts of burden, followed by horses 1,000 years later (3000 BCE).[3] These animals transformed the solar energy stored in grass and other plants into agricultural labor and a means of transport. Of course, certain societies also used humans for this work via the brutal route of slavery and servitude; even this grim business was ultimately powered by the sun, via the food eaten by those forced into labor.*

The sun is also responsible for another important energy source: wind. The sun's rays don't fall evenly upon the planet; some places become warmer than others, and this causes differences in air pressure, creating wind. Humanity has found various ingenious ways to capture this indirect solar energy over the millennia, from sailing ships to windmills. Similarly, it is the sun that drives the evaporation of water, placing water at the top of hills and mountains for it to flow back down again, and (since around 300 BCE) to be occasionally captured by watermills.

These methods of capturing the sun's energy powered human societies for thousands of years, but they all faced strict limits. Wood and canvas could only capture a limited amount of wind and water power, in the right times and places. Only a certain amount of wood could be harvested from local forests each year; only a certain number of animals could be fed per hectare of grazing land. Human labor, too – whether forced or freely given – can only provide a certain amount of energy per day.

All of this changed with the Industrial Revolution. Fossil fuels – which up until then had been used only on a small, local scale – started to be extracted and burned at an industrial level. Coal, oil and gas represent millions of years of handily prepackaged solar energy. The globe-spanning forests of the Carboniferous period spent 40 million years soaking up the sunshine and storing it in their trunks; a few hundred million years later, they form

* While slavery has now thankfully been outlawed across the globe, as many as 29 million people may nonetheless still be in servitude. Enslaved people may no longer be powering the triangular trade of international empires but that doesn't mean that slavery has gone away. globalslaveryindex.org/global-release

the world's great coal deposits. Meanwhile, tiny ocean plants – and the animals that fed upon them – took the energy they'd captured from the sun down to the ocean floor, where millions of years of pressure transformed them into oil and gas.

Suddenly, humanity had access to staggering amounts of stored solar energy – or, rather, the dominant governments and corporations of the 19th and 20th centuries had access to this energy, thanks to their control of the land and military power necessary to secure and exploit these resources. This played a major role in shaping the societies we have today, and control over fossil fuels is still a dominant driving force in global politics.

Of course, those ancient plants and animals didn't just capture sunshine, they also sucked a large amount of carbon dioxide from the atmosphere and locked it safely away underground. This played a large role in shifting the climate from the steamy swamplands of the Carboniferous period to the milder, more temperate and human-friendly climate that we enjoyed up until the middle of the 20th century.

We no longer have that climate. The trillion tonnes of carbon dioxide we've put back into the atmosphere from burning all that coal, oil and gas have heated our planet by almost a full degree Celsius, shifting us into a new climatic reality of increased floods, storms and droughts. If we continue down this fossil-fuelled path, then this will be a century of rising oceans, superstorms, unprecedented drought, collapsing food and freshwater supplies, the probable loss of hundreds of millions of human lives, homes and livelihoods, along with the mass extinction of other species: tragedy on an unimaginable scale.*

Fossil fuels have served their time. The science demands that we leave 80 per cent of conventional oil,

* In my previous book, *The No-Nonsense Guide to Climate Change (2010)*, I wrote a step-by-step explanation of the science behind this, as well as an overview of the history and politics of climate change, and what it will take to avoid the worst effects.

coal and gas (and pretty much all of the 'unconventional' fossil fuels like tar sands and shale gas) in the ground, in order to have a decent chance of avoiding runaway climate disaster.[4] We need to stop burning these deadly stocks of ancient carbonized sunshine as a matter of extreme urgency. That means we need safer, cleaner ways to get the solar energy our societies rely upon.*

Luckily, we've already made a start in this direction. Clean energy technologies have improved enormously in the last few decades. Renewable energy – that is, energy derived from the natural, self-sustaining systems of the planet – currently provides around 19 per cent of global energy use. With the exception of a small amount of geothermal energy from beneath the Earth's surface (see Chapter 4), and tidal energy from the moon's gravitational pull (Chapter 5), all of this energy comes to us fresh from the sun. Some is directly captured in solar generators or heat collectors (Chapters 1 and 4); some is indirectly harnessed via wind and water that has been pushed into motion by the sun's rays (Chapters 2, 3 and 5); and some comes from the solar energy stored inside plants (Chapters 6 and 7).

The sun has always been our main energy source – we just need to choose the best methods for gathering that energy. The great challenge of our lifetimes will be shifting from the stale, poisonous solar energy stored in fossil fuels to the fresher, cleaner sun-powered technologies of the here and now (with a bit of moon and earth power mixed in for good measure). As well as giving us a shot at a safer climate, these technologies have the potential to share access to energy more fairly across the globe, break up the current energy

* Our societies do use *some* energy that didn't originally come from the sun: nuclear power, tidal power and geothermal energy. However, according to the International Energy Agency only 6% of the energy we currently use comes from nuclear power, and less than 0.1% is from tidal and geothermal. So 94% of the energy we currently use ultimately came from the sun.

monopolies and return power to the people in more ways than one. However, they also come with problems and risks of their own; in particular, energy from plants (which makes up half of current global renewable energy use) is currently being pursued in ways that are often unhealthy, unjust, or so destructive as not really to count as renewable at all (see Chapter 6).

This book aims to cut through the jargon and lay out some key information about all of these energy sources in clear, friendly language. In Chapters 1 to 7 we'll look at each form of renewable energy in turn, explain how they work and consider the risks, the challenges, the benefits and the limits of using that technology.

Then, in Chapters 8 to 12, we'll step back and look at the bigger picture. Could we power the whole world with renewable energy? Is the technology good enough, and what changes would be needed in our society to make this possible?

First, though, let's get a few definitions and caveats out of the way.

What do we mean by 'renewable energy'?

The energy sources covered by this book are considered to be 'renewable' because they originate from the sun, the moon and the natural heat within the Earth; so long as these things still exist, we can (in theory) access this energy, as it continues naturally to renew itself.* This stands in contrast to fossil fuels, which are non-renewable within the lifetime of our societies as they would take millions of years to re-form (as well as just the right combination of plants, animals and geological events all over again). Similarly, electricity from current nuclear power stations cannot be counted as renewable energy as this process requires uranium, which is a finite resource.

* The sun should continue to shine at its current level for another 3,000,000,000 years or so, which is probably enough time to be going on with.

Sustainability, fairness, and an admission of bias

It's not possible to write a book that is completely impartial and free of opinion. Everyone has personal biases that affect their writing, and so the best we can do is to be upfront and honest about them.

In this book, I am biased towards creating a world that is sustainable and fair. I want everyone on the planet to

Giga-whats? An explanation of units

The power output of any kind of generator is measured in Watts. This tells us how much electricity the generator could produce if it was running at full capacity in perfect conditions. For example, a typical solar panel might have a peak power output of 250 Watts, which means that, if it's perfectly positioned in full sunshine, it will produce 250 Watts of electrical power. There are 1,000 Watts in a Kilowatt (KW), so our typical solar panel would have a peak output of 0.25 KW.

Measuring the number of Watts tells us how much electricity is being generated (or used) at any given moment. But what happens if we want to add up the total amount of electricity that has been generated, or used, over a specific period of time? Well, that's where we use a unit called the Watt-hour (Wh). If a solar panel produces one Watt of power for an hour, it will have generated 1 Wh in total (which is a tiny amount – it must have been a small or badly placed panel, or maybe it was very cloudy). If a panel produces 250 Watts of electricity for four hours, it will have generated a total of 250 x 4 = 1,000 Wh, or 1 Kilowatt-hour (KWh). The Kilowatt-hour is probably a more familiar measure of energy use, as it's a unit commonly used on electricity bills.

So to recap: Watts (and Kilowatts) are a measure of *power*. They tell us how much electricity is being produced by a generator, or used by an electrical appliance, at any given moment.

Watt-hours and Kilowatt-hours are a measure of *energy*. They tell us how much energy has been generated or used over a given time period.

When we scale things up we start to need larger units. Here are the common standard units for measuring power and energy:

1 Kilowatt (KW) = 1,000 Watts (W): The power needed to run a typical hairdryer.

1 Megawatt (MW) = 1,000,000 W: Enough electricity to keep 1,400 UK homes running at the time of maximum demand (8pm).

1 Gigawatt (GW) = 1,000,000,000 W: The electricity output of a large coal power station at full blast.*

1 Terawatt (TW) = 1,000,000,000,000 W: The amount of electricity

have access to enough energy for a good, healthy, happy and fulfilling life. I want that energy to come from sources that do not threaten the natural systems that we and other species rely on for our survival. I want our energy system to be democratic, and not based on the exploitation of the many for the benefit of the few.

I hope that doesn't sound like an unreasonable bias.

the US could produce if all its power stations, wind turbines and solar panels were generating at full capacity, all at once.

1 Kilowatt-hour (KWh) = 1,000 Watt-hours (Wh): The daily energy consumption of a medium-sized refrigerator.

1 Megawatt-hour (MWh) = 1,000,000 Wh: The amount of energy four typical rooftop solar panels installed in northern Europe might produce in a year.

1 Gigawatt-hour (GWh) = 1,000,000,000 Wh: The annual energy use of 85 average US homes (or 740 average Chinese homes).

1 Terawatt-hour (TWh) = 1,000,000,000,000 Wh: Electricity produced per year by the Yedigoze Dam in Turkey (which is 100 meters high and floods an area of 15 square kilometers).

1 Petawatt-hour (PWh) = 1,000,000,000,000,000 Wh: The amount of electricity used by the whole of India in a year.

To keep things as simple as possible, in this book I will do my best to stick to KW and KWh when talking about day-to-day human-level energy use; GW and GWh (which are a million times bigger) when talking about power stations; and TW and TWh (which are a thousand times bigger again) when talking about global-scale energy use. I'll leave out the Megas and the Petas to try to minimize the number of different units flying around! I'll also try to avoid using KW/GW/TW as much as I can and try to compare everything in KWh/GWh/TWh; and I'll aim to discuss everything on an annual basis (energy use per year) and avoid talking about energy use per day or per month.

There are other ways to measure energy, such as megajoules or 'barrels of oil equivalent', but I'm going to convert everything into the above units, again to avoid confusion and make it easier to compare things.

* For example, Wallerawang power station in New South Wales, Australia. This station used to produce 1GW of electricity when running at full capacity, and burned up to 2.2 million tonnes of coal per year; as of April 2014 it has been closed due to a lack of electricity demand, partly driven by the take-up of rooftop solar panels in Australia.

I just wanted to be clear about where I'm coming from, and how I'm going to judge the different renewable technologies on offer.

But what about nuclear?

Nuclear fission is not a renewable technology, so is not included in this book. If you believe that nuclear energy might have a role to play in a fossil-free future, feel free to add it in to your personal vision of a fossil-free world. However, you may find it doesn't make an enormous difference – according to the International Energy Agency, if governments succeed in their current plans to increase global nuclear power use by 50 per cent (in the face of significant public opposition), this technology would supply just 3 per cent of global energy use in 2035. Nuclear is one of the planet's most unpopular sources of power, with only 38 per cent of people saying they want it to expand (unlike wind and solar, which have global approval ratings of over 90 per cent).[5] As you'll see in this book, we have plenty of other options for generating electricity without using nuclear, so if we're trying to create an inspiring vision for a fossil-free future there's no real reason not to go along with what most people want and keep nuclear stations out of it.

Other possible energy sources – such as nuclear fusion, or sun-powered technology to turn CO_2 into fuel – aren't likely to be up and running in the next few decades (if at all), and so aren't going to be much help in our urgent transition away from fossil fuels. For this reason, they aren't included in this book either (but do go look them up elsewhere if you're interested).

Is this book still relevant?

Renewable technologies are developing fast! It seems as though every week there's a new breakthrough in solar, wind, river or ocean power. To prevent this book from going quickly out of date there'll be plenty of focus on the fundamentals: how does a solar PV cell work? Why is a big wind turbine more efficient than a small one? Why is energy storage important? These are things that should only become more relevant as renewable energy expands.

When it comes to the numbers – how much renewable power can we generate? – you should probably assume that these figures will get bigger as time goes by. However, the urgency of climate change means we can't wait around for the technology to get better. We need to be shifting away from fossil fuels *right now*. So this book also presents a snapshot of the state of renewable energy at the time of writing (March 2015). This technology is already good enough to start a rapid transition away from fossil fuels – and by the time you read this, its potential will be even greater.

1 nin.tl/bottledsunshine **2** practicalaction.org/powering_poverty_reduction **3** nin.tl/plaintexthistory **4** nin.tl/coalunburnable **5** tinyurl.com/mqy6wzd

1 Solar power

'I'd put my money on the sun and solar energy. What a source of power! I hope we don't have to wait until oil and coal run out before we tackle that.'
– Thomas Edison, electricity pioneer, 1931

I'll tell you how the sun rose –
A ribbon at a time.
The steeples swam in amethyst,
The news like squirrels ran.
The hills untied their bonnets,
The bobolinks begun.
Then I said softly to myself,
'That must have been the sun!'
– Emily Dickinson (1830-86)

I'm typing this book on a laptop powered by the sun. The 10 solar panels on the roof of our house generate – over the course of a year – slightly more electricity than we use, with the surplus going back out into the public grid. Our home has become a miniature power station, despite our location in the not-famously-sunny UK.

These days, solar power is serious stuff. Around 200 Gigawatts of solar electricity generators – the equivalent of 800 million of our roof panels – have been installed around the world at the time of writing, and the number is growing rapidly.* Globally, solar power is providing enough electricity to supply every home in France, or California.[1] As more people take up this technology, more money becomes available for researching, improving and developing our methods of solar capture, and the price of the technology falls. For example, the retail price of solar electricity per KWh in the US has

* 139 GW of solar PV and 3 GW of concentrating solar power (CSP) was in place at the end of 2013, according to the European Photovoltaic Industry Association. An estimated further 45 GW of PV was also installed in 2014, along with a few more GW of CSP (detailed figures for 2014 weren't available at the time of writing).

dropped by 95 per cent per cent since 1980,[2] while the price of solar panels in Australia has halved in the last five years.[3] Here in the UK, we put our own personal panels up in 2011, encouraged by a government subsidy.[4] Installing the same 10 panels today would cost us half as much as we paid just four years ago.

One of the major barriers to the expansion of renewable energy has been the artificially low price of fossil fuels. Because oil, coal and gas producers don't have to pay for the damage their products cause to the climate, the market rate for these fuels has always been much lower than it deserves to be. The rest of us are essentially subsidizing fossil fuels, paying with our health, our environment and our shared future.

Despite this unfair subsidy to fossil fuels, the falling price of solar technology means that getting household electricity from solar panels now costs the same as (or less than) getting it from the grid in a number of countries, including Australia, Germany, Italy and Spain[5] – even without government subsidies. The same is already true in 10 US states, and some analysts believe that all 50 states will reach the same position – known as 'grid parity' – by 2016.[6] This could mark a major tipping point, and the beginning of a genuine boom in solar power.

But before we get too excited, let's clear up a few things about solar power. What are the different kinds of solar-powered technology? How much energy can they really generate? What are the limitations, risks and negative impacts of collecting energy direct from the sun?

Electricity from the sun

Photovoltaic power (PV)

This refers to the generation of electricity directly from sunlight. The most familiar form this takes are the gleaming black rectangular panels you might have seen on rooftops like ours. However, PV generators can come in a few other shapes and sizes too.

A single PV generating device is usually called a cell (a typical rooftop panel will contain a number of cells connected together). Most PV cells contain silicon. When light strikes a PV cell, tiny particles of light (called photons) are absorbed by the silicon and converted into a different kind of particle: electrons. PV cells capture these newly produced electrons and convert them into an electric current (electricity is essentially a flowing stream of electrons). This is known as the photovoltaic (PV) effect, which is where the technology gets its name.

Not all the light energy landing on a PV cell gets transformed into electric power. A typical rooftop silicon PV panel will convert between 12 per cent and 18 per cent of the light landing onto it into electricity; this means it has an efficiency of 12-18 per cent.[7] That may not sound like much, but it's good enough to make solar PV a genuinely significant energy producer when it's installed in a sufficiently sunny spot. For example, in June 2014, coal power station owners in Australia found that demand for their electricity was plunging in the middle of each day, thanks to the 3.4 Gigawatts of solar PV installed on 1.2 million buildings across the country. This means that solar power is starting to reduce the amount of coal burned on sunny days in Australia.

At lunchtime on 9 June 2014, Germany generated a record 51 per cent of its electricity from solar panels, while the small island nation of Tokelau now gets 100 per cent of its electricity from PV.

Efficiency vs cost

It is possible to make solar PV with a higher efficiency rating but, generally speaking, the higher the efficiency, the more expensive the technology. The most efficient solar cell in the world (at the time of writing) is made by the company Fraunhofer, and can convert 44.7 per cent of the light falling on it into energy, thanks to some clever improvements called multi-junction and concentrator technology.

Figure 1.1: Roof-mounted PV panels (not ours)

So if it's possible to make higher-efficiency solar cells, why do most domestic and commercial PV installations currently use silicon cells that work at just 12-18 per cent efficiency? The simple answer is cost. These types of solar panel are efficient enough to do the job while being cheap enough to afford. Multi-junction concentrator cells like Fraunhofer's are incredibly expensive to manufacture and so for the moment are only likely to be used in space, rather than on our roofs.

However, if the price of solar technology continues to drop, then more efficient panels will become affordable to more people in more places.

Efficiency vs location

Another important question regarding solar PV panels is: where to put them? They need to be placed in an unshaded spot, ideally facing towards the Equator (so pointing south from the northern hemisphere and vice versa). There are also major advantages in placing panels near where people live, to minimize the costs and energy losses involved in transporting the electricity to where it's needed. This is why rooftops are such a good option, but of course they are limited in number.

One strand of PV technology research is looking into ways of putting solar cells into more locations on existing buildings and infrastructure in cities, towns and villages. Recent innovations include transparent solar cells (that could be installed in windows),[8] spray-on cells,[9] and solar cycle lanes.[10] So far, these technologies have significantly lower efficiencies than standard rooftop panels, but their developers hope they can make them cheap and unobtrusive enough to install in large enough quantities to overcome this problem.

They won't work everywhere, though. For example, a crowdfunding appeal appeared online in 2014, asking people to send money to support a new idea: 'Solar freakin' roadways!'[11] The people behind the appeal wanted funds to research the installation of specially reinforced solar panels into US roads; in just a few months, the appeal raised $2.2 million. However, many energy and engineering experts are unconvinced by the scheme. They point out that roads – which receive a daily pounding by thousands of cars and trucks – are one of the most difficult possible places to install and maintain this kind of technology. The idea of building panels in roadways is a bit like calling for the mass installation of rooftop cycle racks; we can understand the sentiment behind it, but it might make sense to start somewhere else first.

Concentrated solar power (CSP)*

This is another way of generating solar electricity, based on using heat from the sun. It works by using lenses and mirrors to focus the sun's rays onto tanks of liquid (either water or some other form of 'working fluid'). The heat causes the liquid to boil or expand, driving a turbine that generates electricity.

This works well in very sunny places. Most of the

* Also known as concentrating solar power, concentrated solar thermal.

world's CSP has so far been installed in Spain, the US and the United Arab Emirates.

The commonest type of CSP generator in current use is called the 'parabolic trough'. This is – as you might expect – a trough-shaped construction that focuses the sun's rays onto a tube running down its middle. Other forms of CSP include Fresnel reflectors (which use tilted ground-based mirrors to direct sunlight up onto a horizontal tube); Dish Stirlings (which look a bit like satellite dishes made from mirrors, and focus the sun's rays onto a structure protruding from the middle of the dish); and power towers (where banks of mirrors are arranged in a circle around a central tower, also known as a heliostat).

As you might imagine, these are all large and complex structures and so CSP is currently a more expensive technology to build and run than PV.

However, fans of CSP point out that it has one major

Figure 1.2: Clockwise from top, a parabolic trough; a power tower; and a Dish Stirling – all types of CSP plant.

advantage: the ability to keep generating electricity when the sun isn't shining. Some of the newer CSP plants (such as the Solana parabolic trough plant in Arizona) are able to store the liquid they have heated and keep it at a high temperature into the night, allowing them to continue generating electricity long after the sun has set.

Heat and light from the sun

Forty per cent of all natural gas, and a fifth of all coal and oil, is currently used to produce heat.[12] There are various possible ways to replace this fossil-driven heat supply (see Chapter 10), but some of the most straightforward options involve capturing heat directly from the sun.

Solar water heating (SWH), also called solar hot water (SHW)

The most familiar form of solar water heating is probably the roof-mounted solar collector (Figure 1.3). In the simplest version of this device, water flows through the collector and is heated by the sun, and is then transferred into a hot-water tank for storage. In more sophisticated solar water heaters, a specialized 'heat transfer fluid' flows through the collector pipes; this then runs through a heat-exchanging device that transfers the heat from the fluid into the water in the tank. The transfer fluid contains antifreeze and anti-corrosion agents that can prolong the life of the system and reduce the need for maintenance.

Solar water systems can be active, where the water and/or transfer fluid are moved through the system with an electric pump; or they can be passive, where the heat from the sun drives the movement of the liquid through natural expansion. Active systems tend to capture more of the sun's heat and so are more efficient overall; however, they are also more expensive to run and may require more maintenance. Passive systems capture less energy day to day, but are cheaper to install and run.

Domestic solar water systems are typically installed with a gas, electric or wood-fired water heater as a back-up

Figure 1.3: A roof-mounted solar water heater

(usually by adding the solar water equipment to whatever heating system was already in place). The system will use solar-heated water as a preference, but switch to the back-up when the sun-powered hot water runs out.

How well this works obviously depends on how sunny it is. Rooftop solar water panels on a house in the UK will typically provide around a third of the average household's hot water needs over a year. However, solar hot water in countries with more winter sun (such as

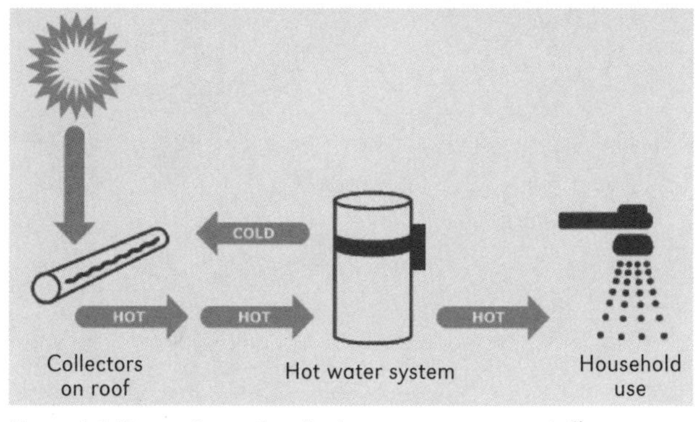

Figure 1.4: How a domestic solar hot water system works[13]

southern Europe, the Middle East, the southern US) can provide around 85-90 per cent of a household's demand.

Some heat-driven industrial processes can also be powered by solar hot water. Industrial solar hot water systems typically use mirrors and lenses to focus the sun's energy, to produce hot water at the necessary temperatures and volumes. They use many of the same techniques as the electricity-generating CSP technology described above – for example, the newly launched solar water installation at Kingsey Falls in Quebec is providing 1,000 KWh of heat per year to a Cascades paper packaging factory, using 1,500 square meters of parabolic troughs.[14]

Solar cooking

Solar cookers work by reflecting and focusing the sun's rays to heat up an oven or a grilling surface, allowing meals to be cooked or water to be sterilized. Solar cookers can range from low-tech devices made from metal foil and cardboard, to higher-end models with mirrors and moving parts that can cook meals for large numbers of people.

Solar cookers need to be left out in clear sunlight

Figure 1.5: A simple solar cooker

for several hours in order to cook a meal, and so they work best in countries where that kind of weather can be relied upon for much of the year. They can provide a good alternative to stoves powered by gas, wood fuel or kerosene, particularly in rural areas with no connection to electricity or gas grids.

Passive solar heating, cooling and lighting

With the right combination of Equator-facing windows, heat-absorbing inner surfaces and the correct building materials, a home or office can soak up heat during the day and then release it slowly at night.[15] For maximum effect, this needs to be part of the building design from the start; however, it is possible to retrofit some buildings to act as solar heat traps in this way, by adding new windows and internal fittings.[16]

In hotter countries, solar-driven cooling systems might also have a useful role to play. These gather energy from the sun and use it to drive a cooling device that can also provide hot water.[17] This technology is still in its early stages and it isn't yet clear how effective (or affordable) it will be.

Good building design can also help to minimize the need for electric lighting, by directing natural sunlight more effectively into home and work spaces during the day.

Depending on the exact location and design, passive solar heating, cooling and lighting combined with good insulation can seriously reduce the amount of energy needed to keep a house at a comfortable temperature. For example, houses built to meet the super-efficient 'Passivhaus' design standard (developed in Germany) only require half as much energy to heat and cool as the average European home.[18]

History and current use

John Perlin's book *Let It Shine* traces humanity's use of solar energy back 6,000 years to neolithic China, where villagers built their houses with south-facing doorways so as better to catch the sun's rays. Both the ancient Greeks and Romans had knowledge of solar building methods, and both the Chinese and Greek civilizations independently discovered how to use solar reflectors to light fires thousands of years ago.

In the 17th and 18th centuries, wealthy Europeans used glass to capture the sun for growing plants, leading to the development of sophisticated greenhouses. The Swiss inventor Horace-Bénédict de Saussure designed a glass-topped 'Hot Box' in 1767 that heated water from the sun's rays – a precursor of modern solar water heating systems.

The use of solar heating and solar hot water progressed sporadically through the 19th and 20th centuries. These technologies had bursts of popularity (for example, in Florida before World War Two), but were superseded by the rise of 'cheap' fossil fuels and the efforts of the coal, gas and oil industries to stamp out their competitors.

The first known solar-powered motors were designed by the French mathematician Augustin Bernard Mouchot in the 1860s. He built the first parabolic trough, and succeeded in using solar power to boil water and drive a steam engine. By the end of the 19th century, US pioneer Aubrey Eneas had built and installed three solar-powered irrigation pumps on Arizona ranches. The machines were six storeys high and weighed four tonnes.

The photovoltaic effect – where the sun hitting particular materials creates an electric charge – was discovered in 1876 by the British scientists William Grylls Adams and Richard Evans Day. These early experiments used selenium rather than silicon, and so only produced a small amount of electric charge.

It wasn't until the 1950s that the US scientists Daryl Chapin, Calvin Fuller and Gerald Pearson discovered the photovoltaic effect in silicon while working on transistors for Bell Laboratories. Their early solar-cell

Something New Under the Sun. It's the Bell Solar Battery, made of thin discs of specially treated silicon, an ingredient of common sand. It converts the sun's rays directly into usable amounts of electricity. Simple and trouble-free. (The storage batteries beside the solar battery store up its electricity for night use.)

Bell System Solar Battery Converts Sun's Rays into Electricity!

Bell Telephone Laboratories invention has great possibilities for telephone service and for all mankind

Ever since Archimedes, men have been searching for the secret of the sun.

For it is known that the same kindly rays that help the flowers and the grains and the fruits to grow also send us almost limitless power. It is nearly as much every three days as in all known reserves of coal, oil and uranium.

If this energy could be put to use – there would be enough to turn every wheel and light every lamp that mankind would ever need.

The dream of ages has been brought closer by the Bell System Solar Battery. It was invented at the Bell Telephone Laboratories after long research and first announced in 1954. Since then its efficiency has been doubled and its usefulness extended.

There's still much to be done before the battery's possibilities in telephony and for other uses are fully developed. But a good and pioneering start has been made.

The progress so far is like the opening of a door through which we can glimpse exciting new things for the future. Great benefits for telephone users and for all mankind may come from this forward step in putting the energy of the sun to practical use.

BELL TELEPHONE SYSTEM 

Figure 1.6: An advert for the 'Bell Solar Battery' from 1956[19]

designs arrived at just the right moment to be picked up and developed by the US military for use in satellites, as part of the Space Race with the USSR.

Between the 1960s and 1980s, solar PV was used for specific applications where access to other forms of electricity was difficult – lighthouses, offshore oil rigs and remote towns and villages. However, it could not yet compete with fossil fuels for most electricity production.

Solar technologies of all kinds were given a boost by a series of spikes in oil prices in the 1970s and the growth of the environmental movement. A million solar water heaters were installed in the US between 1973 and 1986, and seven million were put up in Japan. However, the 1980s saw a drop in oil prices and the rise of politicians like Ronald Reagan and Margaret Thatcher, who set about deregulating the energy industry and promoting the interests of oil, gas and nuclear-energy companies. This made life much easier for the fossil-fuel corporations, while government support for solar power slumped in the US and much of Europe.

Despite this, some governments continued to provide support for solar. A combination of government subsidies and high energy prices led to huge take-up of solar water heating by households in Cyprus and Israel; now 90 per cent of homes in these countries use the sun to heat their water. This technology is also widely used in Austria and Barbados, while around 40 million solar water heaters have been installed in China (representing 80 per cent of the 50 million in current operation worldwide).

Global solar cooker use is hard to gauge precisely; one estimate puts them at 1.5 million,[20] with 600,000 in East Asia and 400,000 in the US.[21]

Solar PV has seen significant growth over the past 10 years, from just a few Gigawatts (GW) of solar panels in 2004 to nearly 140 GW in 2013, and around 190-195

GW by the end of 2014.* This rapid increase included a doubling from 50 to 100 GW between 2011 and 2012. However, this growth has not been evenly spread across the world, with the top 10 countries hosting 84 per cent of all solar PV installations, as shown in Figure 1.7.

This growth seems to have been largely driven by government policies. The governments of Germany, Italy, Japan, Spain, France, Australia and the UK have all used a feed-in tariff scheme to encourage the installation of renewable electricity; this means that anyone who puts up some solar panels (or, in some cases, builds some wind or hydro power too) will then be paid a fixed amount of money per KWh generated, over a given period of time. The Chinese government has been actively subsidizing

Figure 1.7: How much solar photovoltaic (PV) generation has been installed in different countries – a top 10. This top 10 accounts for 84% of all installed PV worldwide. [22]

Country	GW of PV in place at end of 2013
Germany	35.5
China	18.3
Italy	17.6
Japan	13.6
United States	12.0
Spain	5.6
France	4.6
Australia	3.3
Belgium	3.0
United Kingdom	2.9
Rest of world	*22.6*
GLOBAL TOTAL	139

* Detailed and reliable information on solar installations around the world in 2014 is not yet available, so the tables in this chapter use figures from 2013. Something like 45-50 GW of PV and several GW of CSP is believed to have been installed in 2014, so do bear that in mind when perusing the 2013 tables!

solar-power installations, and every US state has some kind of policy to incentivize renewable energy, ranging from tax rebates to cheap loans to state-wide targets.

A number of other countries, including India, Greece and Romania, are all stepping up their PV installations – each of these three installed around 1 GW in 2013, and so could push their way into the top 10 before too long (particularly India, with its large area, high population and sunny weather).

While most of these panels are on the roofs of homes, businesses and community buildings, photovoltaic 'solar farms' are also starting to spring up. These consist of large outdoor arrays mounted on the ground; the US, China, India, Germany and Spain all have more than 1 GW of PV installed this way.

32%
The rate of growth of solar power in 2014.

22%
The annual growth rate required for solar to provide ¼ of global energy by 2040.*

* Assuming the global energy use total laid out in Chapter 9

The biggest of these is the Topaz Solar Farm under construction in California, which will be 0.55 GW when completed (it's currently 0.30 GW). Solar farms make up 15 per cent of global PV.[23]

On top of the 139 GW of solar PV, we can add 3.4 GW of CSP, most of which is currently in Spain and the US. The largest operational CSP plant in the world is the Ivanpah project, also in California, rated at 0.39 GW.

Costs, risks and drawbacks

As we've seen, solar electricity is getting cheaper – cheap enough to compete with standard grid electricity in sunnier parts of the world. However, in less sunny spots (including the UK) it still requires government subsidies to compete financially with fossil-fuelled energy (which, as we've already noted, is artificially cheap because its real impacts on our health and environment aren't being counted).

Even in countries blessed with copious sunshine, solar PV still isn't an easily affordable option for most individuals because the great majority of its costs (80-90 per cent) are incurred at the start of its life. Rooftop solar panels are very cheap to run (especially PV panels with their lack of moving parts) but require an initial investment that can be daunting for the average European, Australian or North American household, let alone residents of the Global South. However, communities around the world are starting to overcome this problem by forming energy co-operatives and fundraising schemes to purchase shared solar panels (see Chapter 11); there are also government schemes that bring solar power to people who need it. Smaller-scale solar devices (such as chargeable solar lighting) are also becoming much more affordable. Meanwhile, solar water heating and solar cookers can be relatively cheap to set up with the right materials and know-how.

Another challenge is the space required for solar generators. The cheapest way to install solar energy is on rooftops, because it requires minimal work and the energy is produced very near to where it's needed; however, there is a limit to the number of Equator-facing rooftops in the world, with both solar water and solar PV competing for that space.

50%

The price of solar panels in Australia has halved in the last five years.

Solar generators can be set up as stand-alone devices, but the amount of space available for this varies from country to country. It is possible for certain crops or grazing animals to share space with solar panels; in sunnier countries, this can even be beneficial for

Future solar?

It seems like hardly a week goes past without the announcement of some exciting new solar-technology breakthrough. We're told that various new forms of solar might be just round the corner: cells that are mega-efficient, or transparent, or super-cheap; solar panels that float on water or sit inside roadways or beam energy back from space. Some of these ideas may materialize while others probably won't, and it's incredibly difficult to predict which of them will actually happen. Watch this space...

crops that require shade.[24] However, in cloudier and/or more densely developed countries this may not be the best way to use agricultural land (see Chapter 10). Deserts have great potential for solar installations, but building in remote areas is more expensive, and the energy generated may need to be transported a long way, which racks up further costs.

Because the 'fuel' for solar generators is sunshine, their lifetime environmental impact is very small compared to fossil fuels. However, they do still have *some* impact – they require energy and materials to manufacture and maintain, and take up space in our urban, rural or wild landscapes. Go to Chapter 8 to see how these impacts of solar power compare with other forms of energy generation.

1 nin.tl/shrinkthatfootprint and EPIA figures of 160 TWh of solar electricity generated globally in 2013. **2** nin.tl/disruptivetechno **3** nin.tl/ozsolarvs-coal **4** The UK's renewable Feed-In Tariff (FIT). **5** nin.tl/eclareon **6** nin.tl/neargridparity **7** nin.tl/pvefficiency **8** nin.tl/windowpower **9** nin.tl/sprayonsolarcells **10** nin.tl/solarcyclelane **11** nin.tl/roadwaysolar **12** nin.tl/baskingheat **13** nin.tl/solarwaterheat **14** nin.tl/kingseyfalls **15** wbdg.org/resources/psheating.php **16** nin.tl/solarretrofit **17** nin.tl/solarcooling **18** passivhaus.org.uk **19** nin.tl/batterysolar **20** nin.tl/solarbox-cooker **21** nin.tl/1E627TZ **22** epia.org/news/publications **23** nin.tl/1Brm4aw **24** nin.tl/1B4WSI5

2 Wind power

'What giants?' asked Sancho Panza.
'Those you see over there,' replied his master, 'with their long arms. Some of them have arms well nigh two leagues in length.'
'Take care, sir,' cried Sancho. 'Those over there are not giants but windmills. Those things that seem to be their arms are sails which, when they are whirled around by the wind, turn the millstone.'
– Don Quixote by Miguel de Cervantes (1605)

'But let's ignore my opinion for a second, and compare wind farms to other, more traditional, forms of energy production: coal, nuclear and oil. I may be biased, but I think wind farms are definitely the least horrific to look at.'
– Tom Pritchard, gizmodo.co.uk

The sun heats some bits of the Earth's atmosphere faster than others. This creates differences in air pressure, and sends air rushing from high-pressure to low-pressure areas. This air movement is better known as wind.

The precise direction and speed of the wind is also affected by the Earth's rotation and geography. Some parts of the planet get steady, predictable winds; elsewhere, the wind can be more changeable or fickle. For large regions of the planet, it is a very promising source of energy.

Wind turbines

These spinning wind-powered electricity generators are an increasingly common sight in many countries, particularly in western Europe, the US, India and China. They vary in size and design, but all work in the same fundamental way: they have blades which turn in the wind, driving an internal generator that creates a flow of electricity.

Most modern turbines spinning today are the familiar tall, sleek design with three wing-like blades.

Figure 2.1: How a wind turbine works

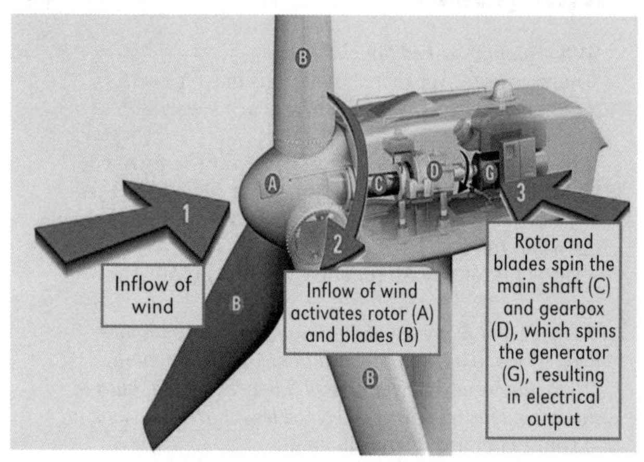

This is largely down to the Danish wind industry (see the history section, below), which led wind-turbine production throughout the 20th century and pushed wind power development down this particular path.[1]

The bigger the wind turbine, and the stronger the wind, the more power can be generated. This isn't a simple relationship – if a wind turbine's blades are doubled in size, it can produce up to four times as much power; if the wind speed doubles then the turbine can produce up to *eight times* as much power. This is why location is so important for wind turbines; building a few large turbines in a windy place will generate a lot more power than building many small turbines in less windy spots.

The key parts of a wind turbine are the *rotor* (the bit with the spinning blades); the *nacelle* (the box behind the blades that contains the internal workings); and the *tower* (the pole that the nacelle sits on).

Wind turbines range greatly in size, from small domestic models a meter in diameter that might (if well placed) generate a few thousand KWh per year, up to giant industrial turbines like the Vestas V164, which has a blade diameter of 164 metres and in a windy offshore spot could generate 35 million KWh per year (35 GWh). This is enough to supply 7,600 UK homes with electricity.

Other uses of wind power

Wind-powered pumps are used to draw water out of the ground in rural communities, for both domestic and agricultural use, in countries such as South Africa and Namibia. Some of these pumps reflect a long history of water-pump use in the area, while others are more modern installations to help provide water in areas not connected to a water or electricity supply.

Sailing ships may also be making a comeback. A number of small shipping companies are experimenting with sailing ships as a fossil-free means of transporting international cargo. Some of these ships are purely wind-powered, others include a back-up for when the wind dies: electric engines charged by solar panels. We'll come back to this form of wind power in Chapter 10.

History and current use

We don't know exactly when a human being first attached a sail to a boat and harnessed the power of the wind. We do know that for most of the last thousand years, knowledge of the world's winds – and access to the ships that could best use them – was vital to trade, exploration and empire-building, until the spread of fossil-fuelled ships in the 19th and 20th centuries brought the age of sail to an end.

Back on land, the ancient Babylonians may have had wind-powered water pumps 3,800 years ago; there were certainly wind-powered machines in Persia by the 10th century. Over the last 1,000 years, many different nations built wind-powered devices. These weren't just the familiar mills for grinding grain; much of the Netherlands was claimed from the North Sea using wind-powered pumps, while farmers around the world used similar devices to drain marshy fields or to irrigate dry ones.

Around a quarter of a million of these early wind-powered machines were built in western Europe alone.[2] They reached their peak between 1750 and 1850, then declined as coal and oil-powered devices became more widespread; however, tens of thousands of windmills were still spinning in Europe at the start of the 20th century, while hundreds of thousands of windpumps were supplying water to US farms, livestock and steam trains.

These windmill and wind-pump designers had already figured out a number of important things about turning the wind into useful energy. Mills were built that could rotate to face the wind, or turn away from it to avoid damage during storms; different sizes and numbers of sails or blades were found to suit different locations and wind speeds.

Engineers have been able to generate electricity from the wind since the late 19th century. With any electric

Figure 2.2: An old windmill next to a modern wind farm in the Netherlands

Figure 2.3: Charles Brush's wind turbine, 1888[3]

technology, you can guarantee that some pioneering Victorian eccentric will have built something in their back garden long before it hit the mainstream; and sure enough, a man called Charles Brush built a fantastic-looking 17-meter-diameter wind turbine in his Ohio back yard in 1888, and used it to charge batteries and power his electric lights for 15 years.

Brush's enormous device could produce about the same as a modern domestic two-meter-wide turbine. Wind-power technology improved slowly over the next 70 years, with most development in the US and Denmark. In 1957, a man called Johannes Juul designed and built a groundbreaking 200-KW turbine at Gedser, on the blustery Danish coast. This turbine included an automatic system that adjusted the working of the rotor in high winds, protecting it from damage and greatly extending the life of the device.

Despite these new developments, by the 1960s the infamous convenience of fossil fuels had driven wind

Figure 2.4: 'Mod-2' two-bladed turbines in Goodnoe Hills, Washington in 1981

energy out of favor in industrialized nations. However, the rise of the environmental movement in the 1970s saw an increase in public interest in renewable power, including wind; a number of governments funded research and development programs, including the UK, Netherlands, Sweden, Germany and Canada. In the US, the Jimmy Carter government declared its support for wind power and triggered a burst of activity in the country – particularly in California, where more than 1 Gigawatt (1,000,000 KW) of wind generators were installed. The first ever 'wind farm' – multiple generators built together in one spot to supply power to the grid – was installed at Crotched Mountain in New Hampshire in 1980, with 20 turbines rated at 30 KW each. However, the arrival of the deeply unsympathetic Ronald Reagan in the White House brought this flurry of wind construction to an end in the 1980s.

As oil prices fell and environmental ideas dropped out of fashion in many industrialized nations, a few northern European countries – particularly Denmark and Germany – continued to slog away at wind-turbine

Figure 2.5 – Total cumulative wind power installed around the world, 2004-2013 (World Wind Energy Association)

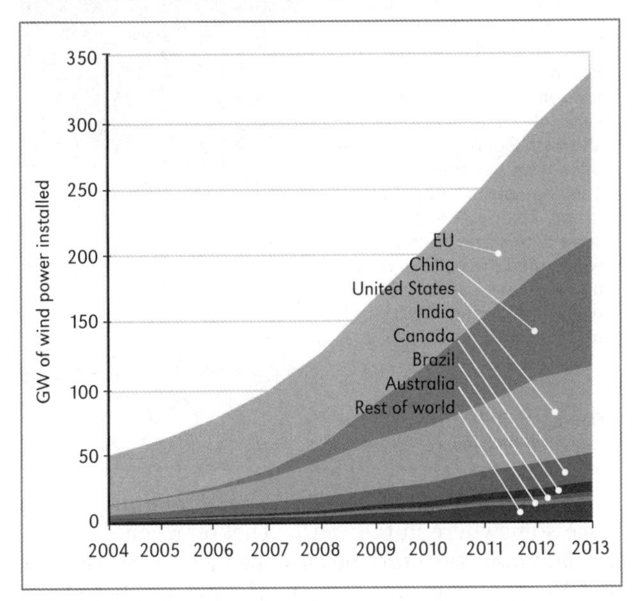

development in the 1980s and early 1990s, largely by themselves. Driven by government support, a range of small- and medium-sized companies tried out various designs, but the most successful proved to be the now-familiar three-bladed design, with a rotor on a horizontal axis facing into the wind. This harks directly back to the 1957 Gedser model, and shows just how pioneering that really was.

As the 1990s progressed, the world gradually woke up to the threat of climate change and wind power began to come back into fashion. The amount of installed wind-power capacity rose from around 2 GW in 1990 to 17 GW in 2000, with Germany, the US, Spain, Denmark and India responsible for most of this growth. This increase accelerated in the 2000s and 2010s, particularly as China got involved, as shown in Figure 2.5, above.

Vertical or horizontal?

Upright turbines with a horizontal axis have become the favored design for churning out the largest amount of power. But vertical- axis designs (as shown in Figure 2.6) have also been tried out over the years. They're often popular in urban settings because they cope better with the turbulent winds found in built-up areas, and are seen as less intrusive than big-bladed horizontal-axis turbines.

Figure 2.6: A 'Darrieus' vertical axis wind turbine

By the end of 2013, 318 GW of wind turbines were in place somewhere in the world. This represents at least 250,000 individual turbines installed globally.* Another 47 GW (approximately 35,000 turbines) may have been installed in 2014.[4]

The biggest single wind farm in the world is the Alta Wind Energy Center in California, which has 490 turbines ranging from 1,500 to 3,000 KW each, and a total capacity of 1.3 GW (1.3 million KW). However, it pales next to the huge network of wind farms currently under construction in western Gansu province in China. Sixty farms are planned with a total of 14,000 turbines. If it goes ahead as planned, the final project may reach 20 GW of power (over 6 GW have been installed so far). This would produce around 39,000 GWh of electricity per year – enough to meet the current annual electricity demand of New Zealand/Aotearoa in full.

* The Global Wind Energy Council estimated that there were 225,000 turbines in place at the end of 2012. The total installed wind capacity grew by 12.6% in 2013, which suggests there are now more than 250,000.

The world's biggest offshore wind farm is the UK's 175-turbine London Array, rated at 0.6 GW. Government permission has now been granted for an offshore farm twice that size, also in the UK (ONE East Anglia), and even larger offshore schemes have been proposed in Sweden and South Korea (each could be up to 2.5 GW in size if they go ahead).

Denmark – unsurprisingly – leads the world on the share of wind in its power supply. In 2013 it derived 32 per cent of its electricity from wind power, and this figure rose to 39 per cent in 2014. Other countries that generated a large chunk of their annual electricity use from wind in 2013 include Portugal (21 per cent), Spain (21 per cent) and Ireland (16 per cent), with Poland (9 per cent) and Germany (8 per cent) not far behind. Wind power was Spain's biggest single source of electricity in 2013, generating slightly more than the country's nuclear plants. On one windy day in December 2013, Denmark's turbines produced 102 per cent of the electricity used by the country that day, the equivalent of powering the entire nation by themselves for 24 hours.*

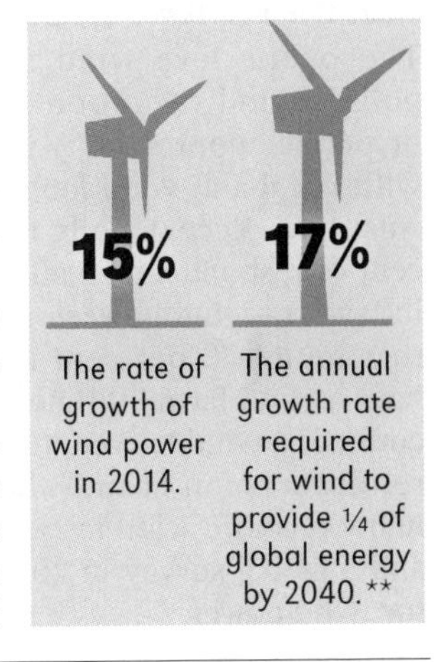

Like solar panels, the costs of wind turbines are dropping as more are being installed around the world. The average price of

* In practice, the electricity from this wind power won't have been produced at exactly the correct times to match the country's demand perfectly – most of the electricity use will have been during the day, and most of the production during the night. So Denmark still needed to keep some other power stations switched on that day, and then either store or sell their surplus wind power to neighboring countries during the night. We'll look more at the issue of matching renewable energy supply with demand in Chapter 9.
** Assuming the global energy use total laid out in Chapter 9.

onshore wind fell 29 per cent between 2008 and 2012.[5] A 2013 Bloomberg New Energy Finance report found that globally, new wind was now cheaper than new gas or coal power;[6] this was even true in coal-producing Australia, where new wind was found to be 14 per cent cheaper than new coal and 18 per cent cheaper than new gas generation per KWh.[7] All of the above figures are raw costs, not including government subsidies or the Australian carbon tax.

Puff and nonsense

There are a number of myths about wind power that its opponents (and, sometimes, its supporters) like to throw around. I'll now try to separate the facts from the bluster.*

Fact: Wind turbines are popular

The public love wind turbines. Seriously. Opinion polls around the world consistently find high levels of public support for wind power, even in countries with vocal anti-wind lobby groups like the UK, US and Australia. Recent polls found that around 70-80 per cent of respondents in all three of these countries were in favor of wind power[8]; in the UK, this included a poll that found 70 per cent of people would be happy to have a wind farm built near their home.[9] In many other countries, wind is even more popular: 80 per cent of EU residents are in favor,[10] there is little opposition to the many onshore wind farms built in Spain and Denmark, and a global survey in 2012 found 93 per cent support for wind power.[11]

* You know those popular myth-busting articles that repeat a series of myths in big bold letters, followed by the reasons why those myths aren't correct? Well, some research has suggested that these articles can be counter-productive, because repeating a myth can reinforce it in people's minds, even if you're only repeating it in order to debunk it! So in this book I've tried actively to counter popular misconceptions about renewable energy *without* printing lots of myths in big letters at the top of paragraphs. I hope it's still interesting and readable this way.

It seems that in certain countries, a noisy minority – often with direct or indirect support from the fossil-fuel industry – have succeeded in slowing the development of wind power, but they are not representative of wider opinion. Wind power is one of the most popular forms of energy in the world.

Fact: Wind turbines don't work when the wind is too low (or too high)
BUT ALSO
Fact: Conventional power stations don't work all the time either

In the words of energy researcher Paul Lynn, 'Perhaps it is unfortunate that wind turbines are so transparently honest.' When a wind turbine is generating at below maximum power it's obvious: the blades don't turn, or only spin slowly. Any passers-by who spot this can then, if they choose, have a good grumble about how useless those pesky wind turbines are.

However, other power sources are intermittent too – they're just less obvious about it. It's a lesser-known fact that fossil-fuel and nuclear-power stations are frequently shut down for repair or maintenance, sometimes at very short notice. This can create a much bigger problem for whoever's managing the power grid. Low or high winds can be predicted pretty accurately a day or two in advance, and planned for accordingly; if a large nuclear station goes offline suddenly, then grid managers have to find a lot of extra generation at very short notice in order to avoid blackouts.

Well-placed wind turbines produce electricity 70 to 85 per cent of the time, but often at below their maximum output; this means they typically produce between 20 and 40 per cent of their potential maximum in any given year. By comparison, fossil-fuel and nuclear plants typically produce between 50 and 80 per cent of their rated maximum each year, due to time spent offline for maintenance and repairs (or due to low demand).[12]

The upshot of all this is that the variability of wind power (and other forms of renewable energy) poses a special challenge to electricity grids, and we'll look at this more closely in Chapter 10. However, electricity grids have successfully dealt with unpredictable shut-downs from other forms of power for decades, and we already have examples of grids functioning well with significant percentages of wind and solar power (such as Germany, Denmark and Spain).

Fact: Wind farms definitely generate useful amounts of energy
BUT ALSO
Fact: Most wind farms can't match the output of a big fossil-fuelled power station

Both of the above are true, which can lead to some confusion! Let me try to explain.

Critics of wind farms like to claim that they produce hardly any energy. By comparing, say, a 100-meter onshore wind turbine (producing around 3 GWh each year) with a large coal-fired power station (producing 10,000 GWh per year), wind-bashers can say 'Aha! Look! You'd need, like, three and a half thousand turbines to replace one power station! Clearly not worth it.'

Now, there's nothing factually wrong with this comparison. The problem is that it's not a particularly useful comparison to make.

A single wind farm is unlikely to match the output of a fossil-fuel power station (unless it's a Gansu-style megafarm). But this shouldn't be surprising – fossil fuels represent an extraordinary concentration of solar power, built up over hundreds of thousands of years. We can't expect to extract that kind of energy from the wind in just a few days, weeks or years – not without a LOT of turbines.

This isn't the fault of wind power. Modern turbines have become far more efficient over the last few decades, and can generate serious amounts of energy, but they're up against physical limits – there's only so much energy that can be extracted from the wind at any one time. Despite this, wind is still one of our best hopes for powering the world without fossil fuels.

And this is the fundamental point here. The fact that coal or gas-fired power stations can generate a lot more energy than a typical wind farm isn't a reason not to build wind farms; it's a reason to seriously rethink the amount of energy we use. Would we need a lot of wind turbines to generate as much power as a fossil-fuelled power station? Yes. Would that take up significantly more land or ocean space than a coal power plant, even if you take into account all the coal mines required to feed it? Yes. Do we need to switch away from fossil fuels anyway? *Yes.* So the challenge is to find a way to still provide everyone on the planet with the energy they need using cleaner but more dispersed energy sources like solar and wind, without relying on the seductive, poisonous, oh-so-convenient charms of fossil fuels.

That's why Chapter 9 asks the question: *how much energy do we really need?* Once we have a rough idea of this, we can then think about the best way to supply that amount of energy cleanly, without trashing the natural life support systems that we all rely on for survival. This is a much more useful comparison than wind turbines versus power stations, which is a bit like saying: 'Hey, why are you wasting time peeling and eating all those annoying oranges? These small, poisonous berries contain just as much Vitamin C, for far less effort!'

39%
The amount of Denmark's electricity generated from wind power in 2014 – a new world record.

Flying or floating wind power?

Standard offshore turbines stand on solid foundations on the ocean floor, and so are difficult to build in waters more than 40 meters deep. This is creating a major challenge for countries like Japan, which are keen on developing more wind but are surrounded by deep ocean waters. A possible solution? Turbines mounted on floating platforms, anchored to the seabed by cables. Prototypes are already generating electricity off various coasts around the world; one promising floating turbine in Portugal has generated 10 GWh since 2012, remaining upright and intact through major storms.[13]

Meanwhile, other designers are attempting to launch wind power into the air. Winds at higher altitudes tend to be stronger and more reliable than surface-level winds, and wind turbines floating high above our heads might be less objectionable to anti-wind-farm campaigners.

Figure 2.7: The Altaeros Buoyant Airborne Turbine prototype

Designs being trialled at the moment include devices such as giant kites; gliders attached to spinning cables; and turbines attached to floating balloon-like structures. Who knows which ones might turn out to be successful?

Costs, risks and drawbacks

The main environmental impacts of wind turbines come from the mining and processing of the materials required to build them. They also, unfortunately, pose a small risk to birds and bats that can crash into the blades, and for this reason they need to be sited very carefully to avoid harming endangered species. However, it's important to keep things in perspective because – surprise, surprise – gas, coal and nuclear power are far more deadly to wildlife. See Chapter 8 for a detailed comparison on this front.

Numerous peer-reviewed studies into wind turbine noise have found very few cases of genuine harm from the sound of the spinning blades.[14] The most modern turbines make hardly any noise at all.

Wind power – like solar and wave power – is variable. It's a wild force that can't be switched on and off at our demand. We'll look at the consequences of this in Chapter 10.

An unfortunate blow

Unless you live alone on a windy hilltop, one of the worst places to put a wind turbine is on the roof of your home. Most domestic roofs are far too sheltered by surrounding buildings and trees for the wind to build up to any kind of useful level.

That didn't stop British Prime Minister David Cameron putting a small turbine on his roof as a demonstration of his green credentials in 2007 (back when he was Leader of the Opposition). The turbine lasted about a week before being taken down again. By 2014, Cameron seemed altogether less keen on wind power; his government made changes to the planning laws making it much easier for local campaigners to block onshore wind developments in the UK (while also pushing through laws to help onshore oil and gas companies charge ahead with fracking in spite of local complaints).

Another tricky thing about wind power is the fact that big turbines in windy places are *much* more efficient than lots of small turbines dotted around in less ideal spots. Big, well-placed turbines can give us the same amount of power as a large number of small turbines while using far fewer resources. However, building giant turbines requires a lot more up-front capital, which means they're more likely to be owned by big companies and governments than ordinary citizens or local communities. This has implications for who controls our energy supply in the future. We'll look at this again in Chapter 11.

1 nin.tl/windturbinehistory **2** Paul A Lynn, *Onshore and Offshore Wind Energy: An Introduction*, Wiley, Hoboken, 2011. **3** nin.tl/brushturbine **4** nin.tl/windenergymarket **5** Bloomberg New Energy Finance. **6** nin.tl/technocosts **7** nin.tl/ozrenewables **8** nin.tl/windenergypolls, nin.tl/windenergysupport, nin.tl/albertasupport, nin.tl/cleanenergyviews **9** nin.tl/windfarmsupport **10** wind-energy-the-facts.org/index-71.html **11** nin.tl/1AGeOme **12** ewea.org/wind-energy-basics/faq/ **13** nin.tl/floatingwindfarms **14** nin.tl/hotairoverwind

3 Hydroelectric power

'Big dams are to a nation's "development" what nuclear bombs are to its military arsenal.'
– Arundhati Roy (1999)

'Never give up; for even rivers some day wash dams away.'
– Arthur Golden

Hydroelectric power is renewable energy at its best, and at its worst. It can provide steady, reliable power to electricity grids or remote communities; it can also devastate entire regions, displace hundreds of thousands of people, wreck ecosystems and add huge amounts of greenhouse gas into the atmosphere.

Hydropower is – like energy crops – a perfect example of how renewable energy done badly can be worse than fossil fuels. However, if done well, it can be a valuable addition to our energy mix, because it's generally a steadier source of energy than those temperamental solar panels and wind turbines.

Hydropower is – once again – solar power in disguise. Heat from the sun evaporates water from oceans, lakes, rivers, soils and vegetation, carrying it up into the air. This water eventually becomes rain, and when that rain falls on hills, mountains, or any other raised area of land it inevitably flows downwards, forming streams and rivers. Hydropower is generated by capturing energy from this flowing water, and transforming it into electricity or other useful work.

Types of hydropower

Nearly all modern hydropower is used to make electricity, and so the terms hydropower and hydroelectric are used pretty interchangeably these days.

This simple description covers a huge range of very different projects, including hydroelectric dams, run-of-the-river generators, and pumped storage.

Hydropower is currently the world's biggest source of renewable electricity.* It produced almost 3,800 TWh of electricity globally in 2013; this compares with 620 TWh of electricity from wind and 169 TWh of electricity from solar that year. Hydropower supplied 20 per cent of global electricity and 4 per cent of all global energy use in 2013.

Hydroelectric dams

Most hydropower currently comes from hydroelectric dams. These dams are built on rivers with a strong, reliable flow, where the water is moving directly from a higher to a lower point (rather than slowly meandering across a plain, for example). The dam blocks the river,

Figure 3.1: The Hoover Dam, US

---

* Note that it's the biggest source of renewable *electricity*, but only the second-biggest source of renewable *energy* overall. The largest source of renewable energy today is wood-burning for heating and cooking; see Chapter 6.

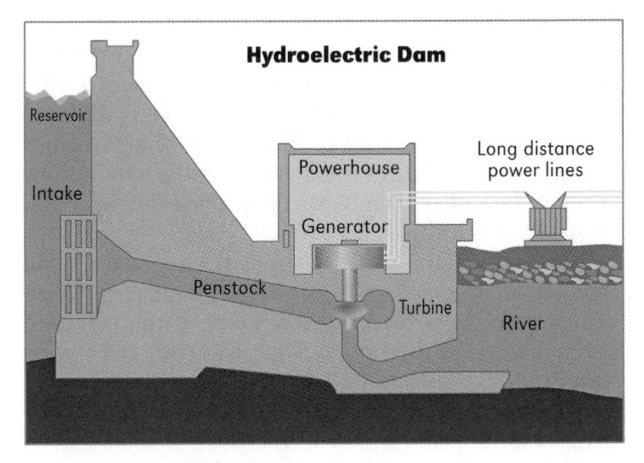

Figure 3.2: How a hydroelectric dam works

creating a large artificial lake (reservoir) behind it. Depending on the size of the river and the shape of the land, very large areas can be flooded by this kind of dam.

Water is allowed to flow through the dam in a controlled manner via special channels or pipes called penstocks. The penstocks carry the water through a turbine, causing it to rotate. This rotation is then used to generate electricity.

The water then flows onwards, back into the original bed of the river. However, large dams often capture enough water – and energy – to significantly alter the speed and volume of the river from that point onwards.

One advantage of this kind of hydropower is that the amount of water flowing through the turbines can be controlled, allowing electricity generation to be ramped up or down according to demand. In times of lower demand, water can be allowed to build up in the reservoir; that water can then be released to flow through the dam and generate extra power at times of higher energy demand.

Run of the river

This form of hydropower doesn't require major changes to the shape and flow of a river or the creation of a large reservoir. Instead, some or all of the river is diverted temporarily through a set of turbines, then returned to its original route. Power is generated by the natural flow of the river, rather than by the controlled flow of water from a reservoir through a dam.

Some run-of-the-river schemes do include a small amount of water storage known as 'pondage'. These storage ponds are placed just upriver of the turbines; water flows into the ponds when the river is running strongly, and then can be let out again when the river is low. This helps to keep the turbines spinning despite changes in the natural flow of the river. A small dam is typically used to ensure that enough water builds up to flow effectively through the turbines, but this dam contains channels that are constantly open to the river so as not to create a permanent reservoir.

Run-of-the-river schemes take up a much smaller area than big hydroelectric dams, but on the other hand they

Figure 3.3: Chief Joseph run-of-the-river hydro scheme, Washington, US

Figure 3.4: A run-of-the-river scheme

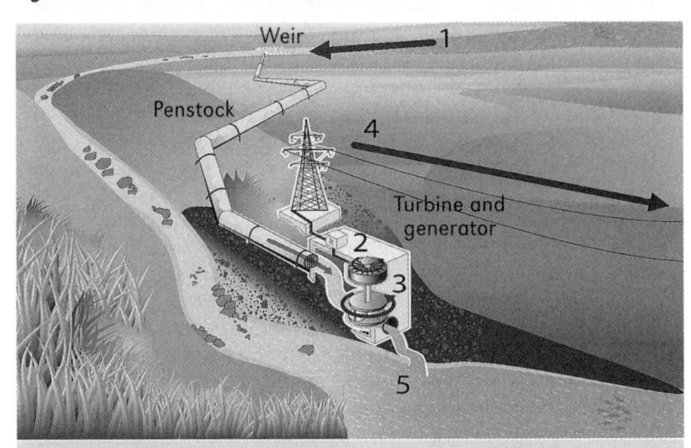

1 Water is diverted from a river into a new channel.
2 The water is passed through turbines.
3 Turbines turn the generator to produce electricity.
4 Electricity is sent to national grid.
5 Water is returned to the river downstream.

are more intermittent and less controllable. The amount of power generated depends on the natural flow of the river, which changes with the weather and the seasons – although it's generally more constant than the wind or the sun.

Pumped storage

This isn't really a form of energy generation – it's a kind of storage – but it's often lumped together with hydropower, so it's worth explaining here.

Ensuring a constant supply of electricity to a national or regional grid is a tricky business. Having a stash of back-up electricity to draw upon is vital for dealing with changes in supply and demand and ensuring the lights stay on. One tried-and-tested method used by many countries is pumped hydro storage. In times of surplus supply, when there's spare electricity flowing into the grid, this extra power can be used to pump water uphill

Community benefit or community destruction?

The differences between big and small hydro projects could hardly be more stark.

In 1979, the Indian government launched a plan to build a network of up to 3,200 dams along the Narmada river, a project that could displace hundreds of thousands of people and destroy farmlands and fisheries. Opposition to the project came to a head in the 1980s and 1990s around the construction of the huge Sardar Sarovar dam; a grassroots protest movement successfully delayed the project for seven years through a series of court challenges backed up by marches, occupations and hunger strikes, and forced the World Bank to withdraw its funding. Despite all of this, the Indian government is now pushing ahead with a plan to increase the height of the Sardar Sarovar dam, in the face of bitter community opposition.

At the other end of the scale, small hydro schemes around the world are providing electricity and/or income to community groups and co-operatives. For example, the Settle Hydro generator on the River Ribble in Yorkshire, UK, generates 60,000-100,000 KWh of electricity each year, most of which is sold to the national grid. The scheme is owned collectively by local residents, and a board of volunteers distributes any financial surplus from the project to local community projects.

into reservoirs. Then, when the grid needs an extra boost of power – due to, say, a spike in demand or a power station going offline – that water can be released to flow back downhill, through turbines that generate electricity.

Non-electric hydropower

Energy from flowing water is still used for some non-electric purposes around the world. Some rural communities use updated watermills to grind grain into flour, or to pump water for irrigation; apparently there are 25,000 such mills operating in Nepal alone.[1]

History and current use[2]

Water power was discovered and developed by the ancient Greeks between the 3rd and 1st centuries BCE. Both the Greeks and the Romans used it for pumping water, either to irrigate crops or to drain mines, and

Figure 3.5: 14th-century Moorish waterwheel (restored) in Cordoba, Spain

used watermills to grind grain and saw wood. The Romans also used a technique called hushing, which used the power of water released from a tank to wash minerals out of rocks and soils.

Water power was also sloshing around in China by the first century of the Common Era, and in India by the fourth century. It was one of the main sources of energy for early machinery all over the world for hundreds of years, along with wind, animal and human muscle power.

A water-powered air compressor called a trompe was developed in Europe in the 1500s, and was in common usage for hundreds of years for pumping air into furnaces, mines and tunnels. The 1700s saw the birth of the Industrial Revolution, and the use of water power to run new devices such as the 'waterframe' spinning machine, developed by Richard Arkwright in Nottingham, UK, in 1769.

Hydropowered devices were still widely used through the 19th century for smaller applications such as pumping air into furnaces or grinding grain; however, the serious power in factories in the industrializing nations came from coal. The black stuff could be transported to wherever it was needed, while hydropower was restricted to factories with a handy river nearby.

However, towards the end of the 1800s hydropower found a promising new use: as a source of electricity. Once again, as with wind power, we find a major early breakthrough being demonstrated by a Victorian enthusiast. Cragside in Northumberland, UK, was the first house in the world to be lit by hydropowered lighting, designed by its owner William George Armstrong in 1878. Around the same time, the first hydroelectric power stations were being built on the US side of Niagara Falls, and in Appleton, Wisconsin. By 1889 there were 200 hydroelectric power plants in the US alone.[3]

The United States remained a world leader in hydropower for the first half of the 20th century, building what was then the world's biggest hydroelectric generator – the Hoover Dam – in 1928 (1.3 GW), and then surpassing it with the Grand Coulee Dam in 1942 (6.8 GW). In the 1940s, 40 per cent of the country's electricity came from hydropower. The US continued to expand hydropower until the 1980s, but other generation sources (coal, gas and nuclear) grew faster, shrinking hydro's share. Today, only seven per cent of US electricity comes from hydro – however, this is still more than all of the country's other renewable sources combined.

Meanwhile, other countries had moved in to steal the megadam crown. In 1984, the Itaipu Dam (14 GW)

Figure 3.6: The Three Gorges Dam, China

opened on the Brazil/Paraguay border. Then, in 2008, the greatest monster of all appeared – China's Three Gorges Dam, with a maximum generation rate of 22.5 GW. Five of the ten biggest hydropower dams in the world are in China.

Several other countries have committed to hydropower in a big way, as can be seen from Figure 3.7.

The 10 countries in Figure 3.7 generate 70 per cent of all the world's hydropower. Over 150 countries use hydropower, but most is concentrated in a few regions, with Asia responsible for about a third. The top five biggest power stations in the world are all hydroelectric dams. Another 26 big dams are under construction, and are set to add a further eight per cent to global hydro generation. At the other end of the scale, tens of thousands of small hydropower schemes of 0.01 GW or less made up only seven per cent of global hydro capacity in 2013.[4]

Figure 3.7: The world's top 10 countries for hydropower

	TWh of electricity generated from hydro in 2013	% of global hydro generation
China	912	24%
Canada	392	10%
Brazil	385	10%
US	272	7%
Russian Federation	181	5%
India	132	3%
Norway	129	3%
Venezuela	84	2%
Japan	82	2%
France	68	2%
Rest of world	*1,145*	*30%*
Global total	3,782	100%

There are 43 countries[5] which source the majority of their electricity from hydro. However, many of these currently have a low demand for electricity that is met by just a few dams.

Costs, risks and drawbacks

When it comes to hydropower, size matters. A 50-KW village hydro scheme to power a few dozen households is clearly a very different proposition from the Three Gorges Dam.

Displacement and destruction

The Three Gorges Dam required the flooding of 1,000 square kilometers of land and the displacement of 1.3 million people. In all, 13 cities, 140 towns and 1,350 villages were submerged, including archeological sites, factories and rubbish dumps. The Itaipu Dam flooded 1,350 square kilometers and displaced 10,000 people. Poorer communities and Indigenous peoples are usually the ones forced to lose their homes, livelihoods and ways of life to clear the way for giant hydro projects. The World Commission on Dams estimated in 2000 that 40-80 million people have been displaced by dams worldwide.

Dams can wipe out entire ecosystems and change local microclimates; they are one of the leading causes of aquatic species extinction.[6]

Megadams like these have faced – and continue to face – huge local opposition before and after their construction. The Narmada Dam project in India has seen decades of high-profile resistance. The successful campaign to stop the damming of China's Leaping Tiger Gorge in the mid-2000s saved a valley of stunning natural beauty, prevented the displacement of 100,000 people and is seen by many as the birth of the modern environmental movement in China.[7] The Ilisu Dam in Turkey was denied international funding following a Europe-wide campaign against its construction.

The notorious 11.2-GW Belo Monte Dam in Brazil

is due for completion in 2015, flooding 1,500 square kilometers of Brazilian rainforest and displacing up to 40,000 people, despite decades of opposition from Indigenous peoples and international NGOs. In Malaysian Borneo, however, the fight is still very much ongoing. Indigenous Kenyah, Kayan and Penan people have blockaded roads and construction sites in protest at plans for 12 dams in the province of Sarawak that would destroy their homes, along with 1,600 square kilometers of rainforest.[8]

Run-of-the-river projects are often presented as a low-impact alternative but, once again, it's all about scale. There are at least 50 examples of run-of-the-river projects bigger than 0.1 GW in operation around the world, with some as large as 3 GW. Even though they don't require the same amount of flooding and disruption as an equivalent 'conventional' hydroelectric dam, once run-of-the-river projects reach this size they inevitably have significant local impacts. Large schemes of this kind can require 'pondage' on a par with a conventional dam (hundreds of square kilometers), and similar knock-on effects.[9] The Bujugali hydro project that has wiped out a series of natural cascades in Uganda and impacted on the homes and livelihoods of 8,700 people is a run-of-the-river scheme.[10]

Even small hydro schemes can have some impact. Redirecting all (or part) of a river or stream through a set of turbines can have an effect on local plants and wildlife, and placing too many generators on the same river can change its flow and have an impact on ecosystems downstream. However, it is much easier to assess, predict and reduce these impacts with small hydro projects, by siting them carefully and running them in a sensitive way.

Water capture

Dams (and bigger run-of-the-river schemes) hold large amounts of water on a semi-permanent basis. This can have a huge downstream impact, shrinking the volume of the river for the rest of its length. This can obviously affect plants and wildlife; it can also reduce the amount of water available to other communities further down the river, for drinking, irrigation or power generation. This is bad enough when it happens within the same country; when dams in one country are restricting access to water in other countries, then it can clearly be a flashpoint for conflict. Tensions are building between the Egyptian and Ethiopian governments over the latter's plans to construct a 6-GW dam on the Blue Nile. Indian and Chinese authorities are similarly at odds over proposed Chinese dam projects on the Yarlung Zangbo river.

Greenhouse gases

Some of you might be thinking: well, it's obviously not great that big hydro projects create these kinds of impacts, but at least they're good for the climate, right?

Sadly, things aren't so simple. While it's true that dams produce far less CO_2 than fossil-fuel plants, we now know that they're pumping out large amounts of a different greenhouse gas: methane.

How? Well, when a dam is built a large area is flooded to create the reservoir. This area will almost certainly contain some trees, grass and/or other vegetation. Once it's been flooded, this vegetation slowly rots under water, transforming into methane, which bubbles to the surface and out into the atmosphere. Hectares of vegetation that had been quietly sucking carbon from the atmosphere are transformed into a source of methane, a greenhouse gas 25 times more powerful than carbon dioxide.

Even when these original plants have rotted away, reservoirs can continue to act as a methane source thanks to organic materials being washed in naturally or dumped by local residents.

How big a problem is this? Well, it obviously depends

on how much plant life was growing in the flooded area, and how much organic matter flows into the reservoir during its lifetime. Limited research has been carried out and only very rough figures are currently available. Smaller hydro schemes in temperate Northern countries seem to have relatively low methane emissions, but large dams in tropical regions may pump out enough methane to make them worse for the climate than fossil fuels.[11] New research suggests that 20 per cent – or more – of humanity's methane emissions could be coming from reservoirs.[12]

Future flows

Climate change is shifting rainfall patterns and melting glaciers. This is already having an impact on the amount of water flowing in the world's rivers; this in turn could reduce the electricity output of hydro schemes. According to International Rivers, sub-Saharan Africa is particularly vulnerable as 60 per cent of the region's electricity comes from hydro. However, the Intergovernmental Panel on Climate Change (IPCC) predicts that the impacts of climate change on overall hydroelectric output between now and 2050 will be small, although it will vary from region to region.[13]

Upstream costs

Like wind and solar power, most of the costs of hydro plants are wrapped up in their construction. Small hydro schemes have proven popular with community groups in Europe and North America, because so long as money can be raised for the initial building the project can then supply a reliable flow of energy – and funds – for decades to come. Big dams, however, are a different matter; their huge construction costs make them the preserve of governments and major corporations. They are often seen as a way for governments and big construction firms to make a lot of money, while local residents suffer the consequences.

Dam and blast?

According to the IPCC, if hydroelectric dams were placed in every technically feasible waterway, we could generate over 14,500 TWh of electricity per year – almost four times as much as current hydro generation, and three-quarters of current global electricity use (18,900 TWh).

But is this a route we really want to go down? Remember, we aren't just trying to power the world renewably for the sake of it – the whole point is to use renewables as a fairer, cleaner, less destructive alternative to fossil fuels. Many large hydro schemes seem to be just as bad – or even worse – than fossil fuels.

This suggests that we shouldn't be calling for a big expansion in hydropower in our quest to replace fossil fuels. That's certainly the line taken by campaigners in Chile, who in 2014 successfully forced the government to cancel five huge dams proposed for the Aysén River in Patagonia, and approve 700 MW of solar and wind generation instead.[14]

In fact, it would probably be a good thing for people, wildlife and the climate if we started tearing big dams down. This isn't as unlikely as it may sound – since the 1930s, the United States has taken down over 1,000 dams in order to restore river ecosystems. This includes the 64-meters-high Glines Canyon Dam on the Elwha River in Washington State, which was breached in 2014 in the world's biggest dam removal so far.[15]

A slow future for hydro?

There's another member of the hydro family that's been somewhat neglected so far: the hydrokinetic turbine. These are devices that are placed directly in a waterway without the need to build a dam or divert a river. They generate less power than other hydro schemes but, unlike the more established methods, they can work in lower-flowing rivers or even canals. This technology is still being improved and developed, but could be a promising source of low-impact energy in the future.

1 crtnepal.org **2** nin.tl/hydrohistory and nin.tl/todayinenergy **3** nin.tl/hydrohistory **4** Calculated from nin.tl/smallhydroworld and ren21.net **5** Albania, Angola, Austria, Belize, Bhutan, Brazil, Burundi, Cameroon, Canada, Central African Republic, Colombia, Democratic Republic of Congo, Costa Rica, Ecuador, Ethiopia, Fiji, Georgia, Ghana, Iceland, Kyrgyzstan, Laos, Latvia, Lesotho, Madagascar, Malawi, Mozambique, Myanmar, Namibia, Nepal, New Zealand/Aotearoa, North Korea, Norway, Paraguay, Peru, Republic of Congo, Rwanda, Switzerland, Tajikistan, Tanzania, Uganda, Uruguay, Venezuela and Zambia. **6** nin.tl/damsextinction **7** nin.tl/savingTigerLeap **8** nin.tl/borneostopdams **9** nin.tl/damslite **10** nin.tl/niledam **11** nin.tl/1AqCxYr **12** nin.tl/methanealarm **13** nin.tl/1MCfdyz **14** internationalrivers.org/node/8338 **15** nin.tl/12dams

4 Heat from earth, air and water

It burst. Suspended, still.
One long vertical sliver
of time nailed into steam and frozen emptiness...
And then it dropped,
spring recoiling into body,
into earth's steam-saline lap,
and tremors shook the hollow crack
as, grumbling and vanishing,
its wild heart juddered back.
– Excerpt of 'Geyser', from *The Night of Akhenaton* by Agnes Nemes Nagy (1922-91), translated by George Szirtes

I want some hot stuff baby this evenin'
Gotta have some hot stuff
– Donna Summer, 'Hot Stuff' (1979)

In this chapter, I'm going to bundle together a few different energy sources.

First, we'll look at geothermal energy produced by physical and nuclear reactions beneath the Earth's surface. This is the first source of renewable energy we've looked at in this book that doesn't ultimately come from the sun.

Then we'll look at those seemingly magical devices called ground-source, air-source and water-source heat pumps, that can extract surplus heat from our immediate environment. This is – usually – heat from the sun, and so heat pumps are another roundabout way of harnessing solar power.

Geothermal power[1]

Once upon a time, a huge cloud of hydrogen and other molecules was floating in space. One day, just over four and a half billion years ago, it began to collapse in on itself. It took a million years or so, but this collapsing cloud became a star – our sun. Other bits of the cloud formed a swirling disc around the star, and over the next few tens of millions of years some of that floating space dust collided, stuck together, and formed planets – including the Earth.

Our home, in other words, was originally formed by bits of space dirt crashing into each other. These collisions transformed the energy from the movement of these bits of dirt (their 'kinetic energy') into heat. So as more dust and particles smashed into the gradually growing planet, the temperature rose, as the energy from the collisions was converted into heat. As the Earth grew bigger, the outer layers pressed down harder and harder on the inner core (thanks to gravity), and increased the temperature still further.

Within 30 million years of the Earth's formation, it had developed into a multi-layered planet with a hot, liquid metal core. The core has been slowly cooling

Figure 4.1: Diagram of the inside of the Earth

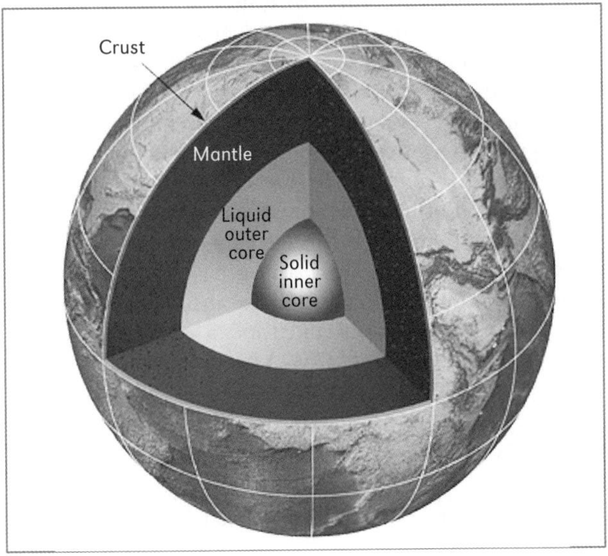

over the last few billion years, creating a solid center; however, the hot liquid layer is still believed to be about 2,200 kilometers thick, with temperatures between 4,000 and 5,700 degrees Celsius.

The next layer up from the liquid core is called the mantle, which is around 2,890 kilometers thick; then the next layer up is the bit we're standing on – the crust – which is 6-10 kilometers thick (below the oceans) and 30-60 kilometers (below the land). However, the liquid core below it all is so hot that it can affect the temperature of the Earth right up to the crust, through almost 3,000 kilometers of planet. The heat from this central core makes up around 40 per cent of the heat used for geothermal power.

So what about the remaining 60 per cent? Well, the Earth's crust contains a number of radioactive elements, such as potassium, rubidium, thorium and uranium. These are gradually breaking down, releasing a constant flow of heat into the Earth's crust.* These natural nuclear reactions are similar to the breakdown that happens within a nuclear power station, but luckily they're happening far below our feet; only a little of the radiation they produce makes its way to the surface, at low enough levels that they pose no threat to our health.

Together, the heat from the core and from these subterranean nuclear breakdowns brings around 390,000 TWh of heat to the surface of the planet every year. This is more than three times our current global energy use. Unfortunately, the great majority of this energy can't be easily collected because it's emerging in highly inhospitable spots (like the bottom of the ocean), or is so thinly spread that we'd need thousands of heat collectors to capture a tiny amount.

However, there are a number of places where this underground heat is concentrated, thanks to natural

* This is thought to be the main source of energy that moves the Earth's tectonic plates, creating earthquakes in the short term and mountains in the longer term.

heat flows beneath the Earth's surface. Some of these spots are very near the surface, and make themselves obvious as volcanoes and hot springs. Others require a bit of digging in order to reach them, but improvements in technology are making more and more of these underground hotspots available to us.

The commonest way to capture this heat for human use is with water. In some areas nature has done the hard work for us: where water flows naturally through an area of high geothermal energy, it will heat up by itself and can then be redirected to heat homes and workplaces. Usually we aren't that lucky, though, and we have to bring our own water; this can be pumped through the hot earth (either at or below the surface), and then used to carry the heat directly to where it's needed.

As well as heating, we can also use this energy to generate electricity. Where steam rises naturally to the surface, it can be used to power a turbine (in a 'dry steam' power station). In locations where the hot water is stuck underground, a shaft can be drilled for the steam to flow up, creating a 'flash steam' power station.*

In places where the water isn't hot enough to form steam, a different liquid (such as butane or pentane) that turns into gas at a lower temperature can be run through pipes into the geothermal hotspot. This second liquid (or 'working fluid') boils and drives the turbine. This is known as a 'binary cycle' generator.

Until recently, all of the above techniques required a significant concentration of geothermal power somewhere near the surface of the Earth, or natural cracks and fissures that allowed water to be pumped down to hotter areas. These conditions are only available in a limited number of places. However, new 'enhanced geothermal' technology should make more of this energy available in more places. This involves drilling

* Hot water below ground can be as hot as 200 degrees Celsius without boiling into steam, because of the high pressure underground. As the water travels up the shaft, the pressure drops and it transforms into steam.

deep enough into the earth's crust to reach previously inaccessible geothermal hotspots. Prototype plants are now up and running in Australia and the US.

Ground-, air- and water-source heat pumps

The idea of geothermal energy is fairly easy to grasp: some rocks are naturally very hot, and we can collect that heat. Ambient heat pumps are different, though. At first glance, they seem like magic; how can seemingly cold earth, air or water be used as a source of heat?

Luckily, it's not too hard to explain (although even after it's explained it still seems rather magical). Anything that's warmer than absolute zero (minus 273.15 degrees Celsius) has *some* heat energy in it. The air, water and soil around us are quite a lot warmer than absolute zero, because they're heated by the sun all day. Air, water and soil temperatures that feel pretty cool to us – say, 5 to 10 degrees Celsius – are warm enough for an efficient heat pump to extract enough energy to heat our homes.

The way the pumps work is slightly different, depending on whether you want to get your energy from air, soil or water. Air-source heat pumps work like a refrigerator in reverse. They use a chemical called a

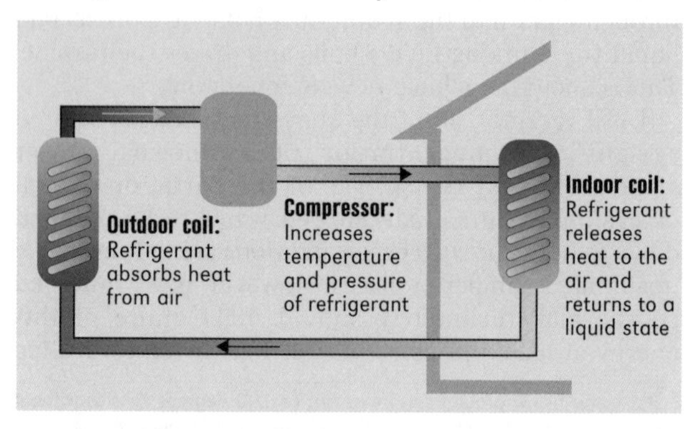

Figure 4.2: An air-source heat pump

'coolant' or 'refrigerant', chosen for its ability to pick up heat easily and boil at a low temperature. This chemical is pumped through a coiled tube in the outside air (inside some kind of ventilated case). The coolant collects heat from the air outside and then flows back into a heat-exchanging device. Inside this device, the heat is transferred from the coolant to where it's needed – the indoor air, or the water in a central heating or hot water system.

Ground-source heating works in a very similar way. However, unlike the air – which flows constantly over the air-source heat pump, bringing new heat with it – the soil tends to stay stubbornly in the same place. This means that in order to be efficient, a ground-source heat pump needs to cover quite a large area in order to collect enough heat. Heating a large house with this technology would typically need 120-180 meters of narrow, coiled underground tubes. Rather than fill these tubes with expensive (and potentially polluting, if it leaked) coolant chemical, most ground-source systems use water with a bit of antifreeze mixed in. This water is pumped through the soil, capturing heat from the earth; it then runs through a heat exchanger that transfers the heat into a second, smaller loop containing a coolant chemical. This

Figure 4.3: Ground-source heat pump coils, during installation

second loop then takes the heat where it needs to go, in the same way as an air-source pump.

Water-source heat pumps can work in the same way as ground-source pumps, or – if the water in question is a large body of water, like a river – water can be pumped directly out of the source, through a heat exchanger and then back to where it came from.

You may have spotted a problem here. All of these systems involve running a pump. That's going to require some energy, in the form of electricity. So a heat pump isn't a straightforward, one-way energy source like a solar panel or a wind turbine – you need to put some energy in, in order to get energy out. However, so long as the heat pump is well-designed and well-positioned, and the temperature of the heat source (air, soil or water) is high enough, then it should generate significantly more energy than it uses.

History and current use

Hot springs have been used by humanity for washing and cooking since prehistoric times.[2] The earliest known purpose-built spa dates from the third century BCE in China; the Romans (of course) used the hot springs at Aquae Sulis (now appropriately known as Bath, in the UK) to provide hot water and underfloor heating for their bath houses in the first century.

The first-known use of hot springs to heat a whole district was in Chaudes-Aigues in France in 1332; it remained in operation until 2009, when the hot water was redirected for use in a public spa. From the 14th to the 19th centuries, geothermal power remained a very localized energy source, used for home heating, public baths and the occasional industrial use (such as the extraction of boric acid from volcanic mud in Italy in 1827). It was only really available in a limited number of places where hot water or steam flowed close to the earth's surface. The end of the 19th and the start of the 20th century saw the development of new district heating systems in the US and Iceland, and the world's first geothermal power station in Italy.

The heat pump was invented by the British physicist William Thomson (better known as Lord Kelvin) in 1852, but at first was mainly used as a cooling rather than a heating technology. When Switzerland faced fuel shortages in the First and Second World Wars, pioneering Swiss engineers experimented with ground-source heating (with little success) but by 1943 had a number of effective water-source heat pumps up and running.[3] A few years later, the US inventor Robert C Webber burnt his hand on the hot outlet at the back of his deep freezer and wondered if he could make the technology work in reverse. By laying copper piping underground, he built what is thought to be the first functioning ground-source heat pump and used it to heat his home.

These technologies spread slowly around the world in the 20th century, but interest in geothermal power and heat pumps picked up after the 1973 oil shock and in response to the unfolding climate crisis in the 1990s. Heat-pump technology was seized on in chilly Sweden

Figure 4.4: Geothermal electricity plants in the top eight countries (in MW of generation capacity, 2013)

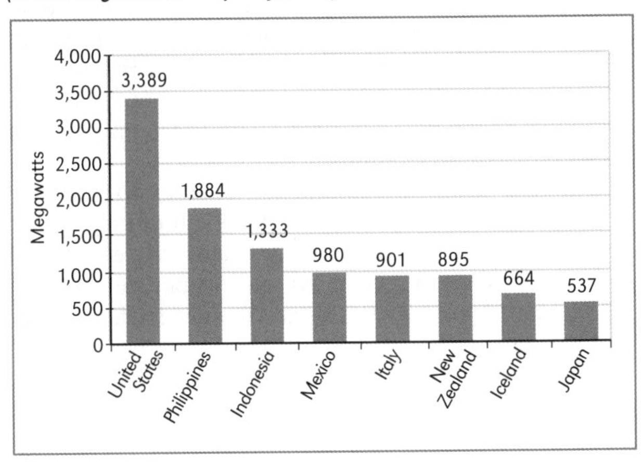

The transformation of Reykjavik

Back in 1900, the capital of Iceland had no electricity, and little heating other than kitchen hearths burning wood, peat and sheep dung. The next hundred years saw an extraordinary transformation. Water from Iceland's abundant hot springs was first used in the 1930s to provide heating for a school, a swimming pool, a hospital and 60 homes. By 1970, direct heat from the rocks and springs was providing 43 per cent of the country's heating needs; today this figure stands at 87 per cent, and includes the heating of greenhouses to increase local food production.

The first electricity plants in Iceland were hydropowered, or ran on expensive imported oil. Experiments with geothermal electricity began in the 1940s, and the first commercial geothermal power station began generating in 1969. Today, geothermal provides 26 per cent of Iceland's electricity, with 74 per cent from hydro and just 0.1 per cent from fossil fuels (largely for back-up purposes).[8] The country is now considering exporting energy, either via undersea electric cables or by using surplus electricity to make hydrogen fuel.[9]

in the 1970s in response to high fuel prices; they became leaders in the field, and now almost half of detached homes in Sweden have an air-source heat pump,[4] while ground- and water-source pumps are also common.

In more recent years, heat-pump technology has begun to take off in other countries too. By 2008, there were 130 million domestic heat pumps in place worldwide, and another 15 million installed in commercial buildings;[5] between 1.5 and 2 million more are being installed each year.[6] The great majority of these are air-source pumps.

The use of hot rocks and hot springs for heating buildings now provides around 90 TWh of energy per year. China is the world's biggest user of geothermal heat, harnessing between 20 and 50 per cent of the global total; other large users are Turkey, Iceland, Japan and Hungary.

Geothermal electricity has grown steadily over the past 40 years, from less than 1 GW of generation capacity globally in 1978 to just under 12 GW in 2013.[7] Currently,

25 countries have geothermal generators in operation, but 90 per cent of the capacity is currently concentrated in just eight countries, as shown in Figure 4.4.

Costs, risks and drawbacks

Geothermal energy isn't available everywhere. New advances in technology are opening up possibilities in more and more places, but certain countries just have more geothermal hotspots than others.

Geothermal plants that bring underground water to the surface (dry steam, flash steam and enhanced plants) also bring a small amount of carbon dioxide and methane up with them from below the earth. However, these are in very small quantities compared with accessing the same energy from fossil fuels (see Chapter 8). Binary cycle plants don't have this problem, because the hot water stays underground.

Geothermal energy requires creating a less-than-beautiful industrial site (see Figure 4.5), and needs some energy to keep it running. However, it requires far less water and land space per KWh than fossil fuels (see Chapter 8).

Figure 4.5: Nesjavellir Geothermal Power Station in Iceland

Interestingly, because it involves bringing up water from beneath the ground – and, in the case of enhanced geothermal, actively pumping water down a borehole – geothermal energy can create some of the same risks as fracking for oil and gas.* The water brought to the surface by flash steam, dry steam and enhanced geothermal sites contains various minerals and materials from beneath the earth, sometimes in quantities that would be harmful to human, animal or plant health. This water is poured or pumped back underground (unlike fracking sites, where used water is transported away as waste) so there is only a low risk of this water spilling or leaking from the plant, but it is still a real concern. There is also a small risk of poisonous underground gases such as hydrogen sulphide reaching the surface. These dangers need to be carefully managed.

Like fracking, there is also a small risk of creating mini-earthquakes from enhanced geothermal plants. The danger is significantly lower than with fracking, however, because a single geothermal drill site can keep producing energy for decades, unlike fracking sites, which tend to be exhausted after a few years, requiring more drilling elsewhere.

It's fair to say that despite a few similarities, enhanced geothermal is a cleaner and safer process than fracking – but then, to be honest, most things are.

Another concern is that many of the world's volcanoes are important religious or spiritual sites, particularly for Indigenous peoples. Any industrial development in these areas should only be carried out with the active consent and involvement of local communities, for whom the area could have serious cultural significance.

---

* Fracking (the snappier name for horizontal slickwater hydraulic fracturing) involves drilling into rock formations that contain pockets of oil or natural gas; a high-pressure mixture of water, sand and chemicals is pumped down the drill-hole and through cracks in the rock to force out the fossil fuel. This is a highly controversial process that is facing fierce opposition around the world. See nin.tl/frackfiles

Heat pumps are considered to be one of the cleanest and safest forms of renewable energy. Air- and water-source heat pumps are cheaper than ground-source ones, but the efficiency of air source varies with the weather, and water source is only suitable for specific sites (next to sizeable rivers, lakes or ponds). Air source has also come under fire in some countries, particularly the UK where installed systems have turned out to be far less efficient than promised. However, air-source pumps set up in similar conditions in Germany worked just fine, leading the UK Energy Saving Trust to conclude that many UK pumps had been wrongly sized, set up incorrectly and run badly, so that this was the fault of the installers rather than of the technology.[10]

Ground-source heat pumps are more reliable, but are disruptive to install and need lots of space. A typical system might require the digging up of 400 square meters per household, to a depth of several meters. There is also a limit to the number of homes that could be fully heated by ground-source pumps in a densely populated area, because the underground heat only recharges slowly and if too many people are drawing on it at once it will run out. In these scenarios, a combination of different heating sources (ground source, air source and solar water heating, for example) might work best.[11]

Heat pumps are currently expensive to install, but should – if properly set up – pay for themselves in saved energy costs over their lifetimes. If demand for them rises, then they should become more affordable, as is currently happening with solar panels.

Energy currents

A water-source heat pump has just been installed at Kingston-upon-Thames in the UK. It should suck enough heat from the River Thames to power 150 homes and a large hotel. The UK government released a report in 2014 mapping out 40 rivers and estuaries that could potentially provide enough energy to heat 20,000 of the country's homes.[12]

Geothermal/solar hybrids

Experiments are under way to see if these two technologies could be combined into one super-renewable. Power plants could use hot rocks from below and the sun from above as sources of heat to drive liquid through a turbine and generate electricity, in a more efficient way than either of these sources could manage on their own.

Heat pumps also require some energy input at the point of use, so require a supply of renewable electricity in order to be considered as properly clean technology.

1 William E Glassley, *Geothermal Energy: Renewable Energy and the Environment*, CRC Press, 2014; Ernst Huenges (ed), *Geothermal Energy Systems: Exploration, Development and Utilization*, Wiley, 2010; Ronald DiPippo, *Geothermal Power Plants: Principles, Applications, Case Studies and Environmental Impact*, Elsevier, 2012; nin.tl/1EmvTX2 **2** nin.tl/1Emw1FT **3** nin.tl/zoggeng **4** nin.tl/1G7joyH **5** nin.tl/1EgvC5v **6** nin.tl/heatpump-market **7** nin.tl/1GRVjz6 **8** nin.tl/energyiceland **9** nin.tl/icelandgeo-thermal **10** nin.tl/1MCKEHh **11** nin.tl/1FjwF8K **12** nin.tl/thamestyneheat

5 Wave and tidal power

There is a tide in the affairs of men.
Which, taken at the flood, leads on to fortune.
– *Julius Caesar* by William Shakespeare

'If you're having a bad day, catch a wave.'
– Frosty Hesson

This is one of the shorter chapters – not because wave and tidal power aren't useful forms of energy, but because they're still in the early stages of development. They are currently only providing a tiny amount of global energy and the technologies involved are generally considered to be in their infancies.

I love the fact that tidal power is essentially moon power. The gravitational pull of the moon drags the oceans in different directions depending on the moon's exact location in relation to the Earth. This creates strong, predictable ocean movements that we know as tides.

Wave power, on the other hand, is a form of wind power, which is in turn driven by the sun; so wave power is, in a roundabout way, yet another form of solar power. Waves are created by winds striking the ocean's surface; the power in the waves depends on the strength of the wind, and also the distance travelled by the wave (called the fetch) and other factors.

Why have wave and tidal power fallen behind wind and solar in the renewables development race? Well, it's far more difficult to build things in the ocean than on the land. Any sea-based technology has to withstand salt water, waves, tides and storms, and any electricity generated has to be transported back to the shore. In addition, unlike a river, the ocean doesn't flow conveniently in a straight line; it's a huge, shifting, complicated, powerful mass of water that moves, swells and recedes in all kinds of directions at once. Capturing its energy requires some particularly tricky technology, so it's probably not surprising that a lot of attention and investment has gone into other forms of renewable power first.

However, wave and tidal power – or 'marine renewables', as they're sometimes called – are finally getting more of the attention they deserve. A number of different devices are at various stages of testing and deployment, each using a slightly different method to capture the energy of the ocean. At this stage, it's hard to say which is most likely to succeed. I'll introduce the main contenders briefly here and then we'll all just have to watch and see which ones turn out to be the most useful!

There are certain areas of the globe where the waves and the tides are particularly powerful and harvestable. Interestingly, they tend not to overlap; most of the places that receive a lot of tidal energy aren't the best places for wave power, and vice versa. Figures 5.1 and 5.2 show the best spots for tidal and wave power around the world; note that the UK is a rare example of a country well supplied with both waves and tides.

Types of wave and tidal power[1]

Wave power generators

Any surfer will tell you that there's a lot of power in waves. Unfortunately, waves are also notoriously messy.

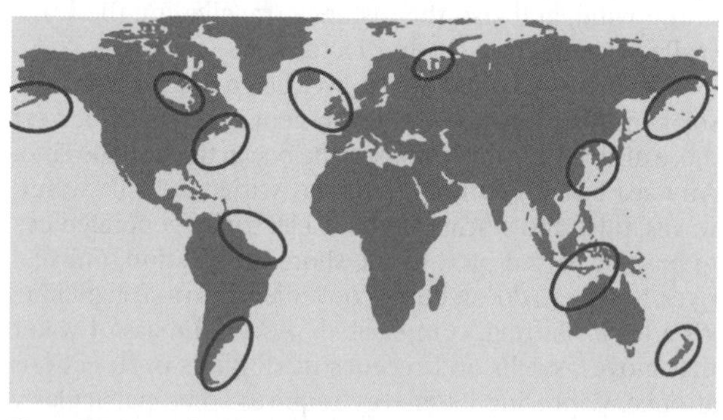

Figure 5.1: Map of tidal hotspots

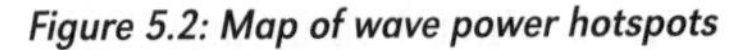

Figure 5.2: Map of wave power hotspots

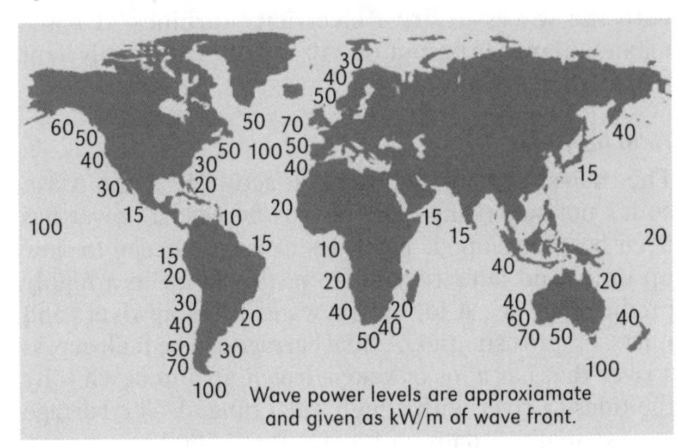

Rather than rolling across the sea in neat predictable lines, they frequently change their height, direction and frequency in response to a number of different factors. A gentle ocean breeze will produce regular waves, but a storm will send waves of different sizes in multiple directions. The height and direction of the waves can then be altered by their interactions with rocks, islands and shorelines, or by changes in the depth of the ocean.

All of this poses a real technical challenge to any would-be designer of wave-power technology. How on Earth (or sea) can these unpredictable, changeable watery beasts be harnessed to provide useful energy? The European Marine Energy Centre (EMEC) describes eight different types of wave-energy converter (WEC) that are currently being trialled around the world.* These range from floating snake-like structures to giant seabed paddles.

The world's first grid-connected wave-power generator was switched on in Perth, Australia, in February 2015. It consists of a series of large submerged

* You can see some nice animations of these designs on the EMEC website: emec.org.uk/marine-energy/wave-devices

buoys, tethered to the sea bed, which bob up and down with the waves to drive electricity turbines. It has a maximum power output of 240 KW, roughly equivalent to a 45-meter wind turbine.

Tidal barrages

The easiest place to see tides in action is at the coast, so it's not surprising that this is where tidal power has been best developed. The tides cause the ocean to flow up onto, and later recede from, the shore in a highly predictable way. A lot of this water flows up rivers and inlets, or into estuaries. A tidal barrage can be built across a river that has a lot of water driven up and down it by the tides (a river with a high 'tidal range'). The barrage is, essentially, a dam with gates that is built across the river. The barrage captures water as the tide brings it up the river, and stores it in a reservoir. A sluice gate opens for the water to flow into the reservoir at high tide, and is then closed before the water can run out again. The water from the reservoir can then be released to flow back to the sea through pipes or channels containing turbines, generating electricity as it goes.

Tidal lagoons

A less disruptive option than a tidal barrage, a tidal lagoon can be created by 'walling off' a section of sea in a coastal area with a large tidal range. The wall keeps the lagoon separate from the surrounding ocean, apart from a series of underwater sluice gates. These gates are kept closed while the tide is going out, keeping the lagoon full of water as the surrounding sea level falls. At low tide, the gates are opened and water flows out of the lagoon into the sea, turning turbines as it goes. Then the gates are closed again, and the process works in reverse: the tide comes in, raising the level of the surrounding ocean while the water in the lagoon stays low. Once high tide is reached, the gates are opened and water flows *into* the lagoon, generating more electricity and returning the

Figure 5.3: Artist's impression of the planned tidal lagoon in Swansea, UK

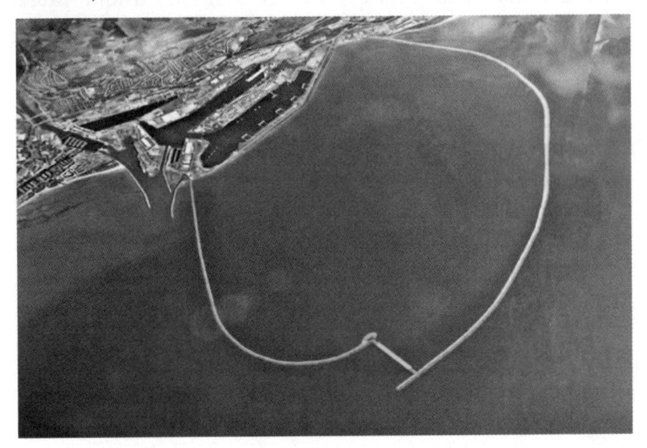

lagoon back to its initial state, ready to do it all over again with the next tide.

Tidal stream generators

Most of the movement of the tides happens under the water where we can't easily see it. The gravitational pull of the moon drags bits of the ocean in certain directions at certain times, causing water to flow in from elsewhere to fill the gaps. When these flows run through narrow channels or straits, or are diverted around land masses, they can be concentrated and focused into strong 'tidal streams'.

Tidal streams can be predicted fairly accurately, but it's a complex business; they can be affected by hundreds of factors, including slight variations in the orbits of the moon and the Earth, interactions with differently shaped bits of coastline and variations in the ocean's depth. The strongest streams aren't always found in the places with the highest and lowest tides; sometimes strong tidal streams flow back and forth against each other in the ocean without reaching a shoreline.

It is possible to harness these streams with turbines on the ocean floor. These act almost like underwater wind turbines, but produce electricity in a much more predictable way.

History and current use

Tidal power has been harnessed for over a thousand years using 'tidal mills'. These work by capturing water in a pool at high tide then running the water back down to the sea through a waterwheel at low tide. The wheel would be used to grind grain or do other mechanical work. The earliest known example was built at a monastery in County Down, Northern Ireland, around 620 CE.

Tidal mills could only be built in certain spots and so were never particularly widespread. By the 20th century they had mostly fallen out of favor, to be replaced by more constant forms of power generation.

There was renewed interest in tidal power as a source of electricity in the 1960s, with the opening of La Rance tidal barrage in Brittany, France, in 1967. With a maximum output of 240 MW, this barrage still provides 540 GWh of electricity every year. Over the next few decades, three similar (but much smaller) tidal stations were built in Russia (Kislaya Guba), Canada (Annapolis Royal) and China (Jiangxia). However, the Rance barrage remained the world's largest tidal power station until 2011, when a 254-MW barrage was constructed in Sihwa Lake, South Korea. This remains the biggest tidal generator in the world, producing around 550 GWh per year.

The five stations mentioned above represent all of the tidal barrage power installed anywhere in the world – a total of just over 500 MW, producing maybe 1.2 TWh of electricity per year. South Korea is planning two more very large schemes, which would add another 2,100 MW just by themselves.

Tidal-stream generators have been installed in South Korea (1 MW at Uldolmok with plans to expand to 90 MW) and Northern Ireland (1.2 MW at Strangford

Lough). Further schemes are in various stages of development in Scotland (10 MW at Islay, 398 MW in Pentland Firth), India (250 MW in the Gulf of Kutch), Canada (20 MW in the Bay of Fundy) and the US (1 MW in New York City).

The world's first tidal lagoon is planned for Swansea Bay in Wales, with a maximum generation capacity of 320 MW. Its developers believe it will produce just under 500 GWh of electricity per year.

The first known design for a wave-energy generator came from one M Girard and his son in 1799. They proposed a device that would use the motion of French warships bobbing up and down in the bay to move long wooden beams, driving mechanical pumps and saws. Sadly, the idea was never put into practice, perhaps because the French navy had other uses for its warships at the time.

Various other ideas for harnessing the waves were tried out over the following centuries. Pioneering ideas included P Wright's 'wave motor', patented in the US in 1898. This used a float, tethered just offshore, to move an onshore generator up and down. Some prototypes were built in southern California but were then abandoned.

M Bochaux-Praceique was more successful in 1910 with the 1-KW wave-powered generation system he built at his home in Royan, near Bordeaux, France. He drilled a shaft down through the cliffs near his home, with an outlet at the bottom to let the sea water in and a turbine at the top which was spun by the air flowing in and out of the shaft as the waves pushed water in and out at the bottom. It did power the lights in his home, but was a *lot* of engineering and digging for – at most – 1 KW of power, which is probably why the idea didn't catch on.

Between the 1940s and 1960s, Japanese inventor Yoshio Masuda experimented with wave power, developing a floating device that could drive a piston to generate a small amount of electricity. This was eventually used to power hundreds of offshore navigation lights, but generated too little power to be used for much else.

It was, as with so many other renewable energy sources, the 'oil shocks' of the 1970s that spurred an increase in wave-power development. A number of technologies were pioneered then that form the basis of many of the designs being trialled today.

Costs, risks and drawbacks

There are good reasons why tidal barrages haven't become widespread, despite La Rance having been operational for almost 50 years. They have many of the same drawbacks as big dams (see Chapter 3) – they involve huge upfront costs, and need to flood a large area to create a reservoir. Coastal and estuary ecosystems are often particularly vulnerable to change, and whole ecosystems can be disrupted or wiped out by these schemes.

These problems explain why the huge proposed Severn Barrage scheme in the UK has never materialized, despite bursts of political support and enthusiasm for the project over the years. While proponents claim that the barrage could provide five per cent of the country's electricity demand, the price has been estimated at between £10 billion and £34 billion ($16-54 billion), and the project faces fierce criticism from environmental groups over the impacts on wildlife.

Tidal lagoons and tidal stream turbines are now generally seen as the future of moon power. They can still disturb wildlife and so need to be carefully sited and monitored; however, they are far less disruptive than barrage systems. Tidal stream turbines have another advantage: like wind turbines, they can be added gradually to a site rather than needing one enormous installation in one go. Building turbines in batches allows costs to be spread out, risks to be lowered and means that lessons can be learned and improvements made as the project unfolds.

As the youngest and least-tested renewable technology, wave power faces major cost challenges at the moment. Once a few successful schemes are up and generating, funding should be easier to find and the technology should start to improve more rapidly. The environmental risks of wave power are similar to tidal-stream technologies: they could disturb marine wildlife and so need very careful siting and monitoring.

1 Paul Lynn, *Electricity from Wave and Tide*, Wiley, 2013.

6 Fuel crops

'The Dakatcha Woodland is a vital area for rare birds, animals and for people in Kenya, but it is under threat from a proposed biofuel plantation... if we didn't have a strong, campaigning civil society in Kenya, this development would have happened by now and this forest would be gone forever.'
– Serah Munguti, Nature Kenya[1]

'Buy land. They ain't making any more of the stuff.'
– Will Rogers

If you want an example of how *not* to do renewable energy, you only have to look at the recent rush to turn plants into fuel.

It seems like quite a nice idea, if you don't think about it too hard. Why not swap some of those dirty coal mines and oil wells for fields of fuel crops, powered by the sun, delivering clean, renewable energy year after year?

Sadly, as soon as you *do* start thinking about it, you realize that things aren't so simple. For a start, we'd need to turn around 600 million hectares of cropland – almost half of global food production – over to making liquid fuels just to power the world's current vehicle fleet.[2] This would clearly be disastrous. Despite this, over the last 20 years, the dream of a plant-powered world has grown from a green-tinted vision to a globe-spanning industry. This has created all kinds of problems, which we'll examine in a moment.

At the same time, it seems that making *some* energy from plants could be really useful – maybe even vital – if we want to shift completely away from fossil fuels. As we'll see in Chapter 10, energy crops could fill important niche roles in powering certain vehicles and industrial processes, and helping to balance electricity grids. Could any of the existing forms of fuel crops fit the bill?

Types of fuel crops

Fuel made from crops is commonly referred to as 'biofuel'. This has become a bit of a catch-all term; some people use it to refer specifically to liquid fuels, but it can be used more broadly to include solid and gaseous fuels as well.

There is also some slightly confusing overlap between biofuels made from new crops, and biofuels made from waste plant (and animal) materials. Getting energy from a piece of waste wood unsurprisingly uses a lot of the same technology and processes as getting fuel from a piece of freshly grown wood, but the environmental and social consequences can be quite different. The same is true for fuels made from organic wastes compared with fuels made from newly harvested food crops. Despite their different origins, fuels made from waste wood, virgin timber, food waste and new crops are sometimes, confusingly, all referred to as 'biofuel' (or biodiesel, biomethane, or biogas, depending on the exact final form of the fuel). The word 'bioenergy' can also be used as an overarching term for energy derived from crops and organic waste.

We'll look at energy from waste in Chapter 7, so this chapter will focus exclusively on 'fresh' bioenergy made from new-grown plants.

Solid fuels ('biomass') for heat

When people talk about 'biomass' as a fuel, they mean solid plant material. The most obvious and widespread of these is wood: the original biofuel, as used in the discovery of fire and a firm favorite for millions of years. Biomass can also refer to fuel made from grasses and plant stems.

Around 65 per cent of global biofuel use is locally harvested wood, charcoal and dung. Around 38 per cent of the world's population rely on this kind of energy for heating and cooking, often burning it on inefficient and polluting indoor stoves.

Figure 6.1: Global bioenergy use 2013. This represents 73 per cent of all 'renewable' energy use, and 13 per cent of total global energy use in 2013

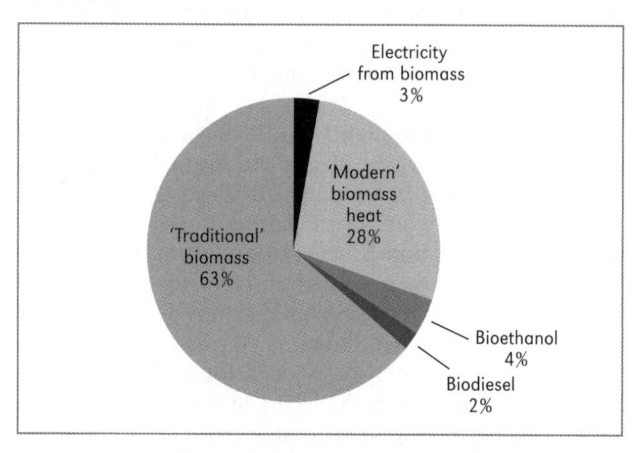

'Traditional' vs 'Modern' wood-burning

While it is important to distinguish between wood that's burned on open indoor fires (which kills around 1.5 million people per year) and wood that's burned in a well-designed and less polluting efficient stove, the way that the words 'traditional' and 'modern' are used by bodies like the International Energy Agency doesn't quite match up with this definition. 'Traditional' is often used as a catch-all term for wood-burning by households in the Global South, while 'modern' refers to all other biomass use. This rather crude definition may give the impression that people in the Majority World using wood fuel in their homes are backward and out of step with the times (even if they collect wood sustainably and burn it in efficient, non-polluting stoves), while households and businesses with wood-burners in industrialized nations are smart and up-to-date (even if they burn imported wood from vanishing forests). In reality, there are examples of both sustainable and destructive wood use in both Northern and Southern communities.

Bearing all of this in mind, good wood stove design can make a big difference. An up-to-date, well-designed wood burner can heat a home using a quarter of the fuel that an open fire would devour, and direct practically all of the smoke safely out of the building.

Electricity from biomass

Some countries are also starting to burn biomass to generate electricity. This usually involves burning wood (or agricultural waste, such as the cane fibres left from sugar production in Brazil) in power stations, in a similar way to fossil fuels. The heat is used to generate steam and drive a turbine, producing electricity.

Ethanol and biodiesel from food crops

Ethanol can be made from sugary or starchy crops such as corn, sugar cane, wheat, sugar beet and barley. It's the same chemical as found in alcoholic drinks and is used for a variety of other industrial processes too. It can also be made from fossil fuels, so when ethanol is made from plants, it's sometimes referred to as 'bioethanol'. Biodiesel is made from oily crops such as soybean, rapeseed, oilseed and oil palm.

Both of these fuels have been presented as possible alternatives to fossil-based liquid fuels (particularly petrol/gasoline and diesel). Ethanol and biodiesel made from food crops (or rather, crops that could have been used for food if they weren't made into biofuel instead) represent almost all of the liquid biofuels in use today. They are sometimes referred to as 'first generation' biofuels, because they were the first to go into mass production in the current wave of biofuel use.

Why use food crops? Well, by their very nature these crops contain a lot of easily accessible energy, in the form of starches, sugars and oils; they're also something that we already know how to grow in large quantities. It's much easier (and therefore more profitable) for fuel companies to link up with existing agribusinesses and divert existing crops into fuel production, rather than starting the whole process from scratch.

Liquid biofuel from wood and grasses

One of the major criticisms levelled at 'first generation' biofuels is that they use crops and farmland that would otherwise be used for food. In response to this, interest has been growing in so-called 'second generation' biofuels made not from the oils and sugars in grains, fruits and vegetables, but from the cellulose in stems and leaves.

Cellulose is a natural starch found in the walls of plant cells. It forms tough fibres that give plants their structure, and is a major component of wood. It's trickier to process into fuel than the sugars, starches and oils found in food, but there's a lot more of it available per plant. Making ethanol from cellulose would allow us to use fast-growing non-food crops like woods and grasses as a source of liquid energy. These plants could in theory be grown on lower-quality 'marginal land' rather than the high-quality farmland needed for first-generation fuels.

The lines are blurry, though; second-generation biofuels can also be made using agricultural waste – such as the inedible stems and husks of food crops – as a source of cellulose for making liquid fuel.

History and current use

Wood is, of course, the longest-serving fuel in human history. It's also possible to argue that the horses, oxen, llamas and other beasts of burden used by humanity over the centuries were a form of bioenergy, with the animals transforming grass and other feed into energy for transport, farming and other purposes.

In the 1700s, whale oil was one of the world's most important fuels, mainly used in lanterns for lighting homes and workplaces. Whaling was big business and global whale populations were in decline. Ironically, the emergence of fossil oil in the 1800s as a cheaper alternative probably saved many whale populations, even as it began the process of global warming (and acidification) that is threatening the oceans today.

Early internal combustion engines in the 1820s were designed to run on ethanol and turpentine. The original Model T Ford in 1908 ran on ethanol. However, the widespread availability of fossil oil in the 1900s meant that diesel and gasoline rapidly drove these plant-based designs into redundancy.

The recent biofuel boom has its roots in the 1990s, when growing public awareness of climate change and oil scarcity spurred interest in alternative fuels. Some environmental enthusiasts blazed a greasy trail by converting their cars to run on waste cooking oil (see Chapter 7), and campaigning NGOs began referring to energy crops as a possible 'transitional' fuel.[3]

Because plants suck carbon out of the air as they grow and then release it when they burn, it was first assumed that the greenhouse-gas emissions from a fuel crop would pretty much balance themselves out. However, this doesn't take account of all the chemical and energy inputs to the crop, and the effects of deforestation and lost soil carbon; once these things are included, the carbon savings from biofuels are small at best, and in many cases they are worse for the climate than fossil fuels.[4]

As these figures were revealed by new research, along with the effect of biofuels on livelihoods and food supplies (see below), most environmental campaigners quickly realized their mistake and withdrew their support – but by this time, big energy and agricultural companies had spotted the potential of biofuels as a profitable new income stream.

Oil companies like BP and Shell, keen to look green without actually changing their core business practices, dabbled with solar and wind power in the 1990s and 2000s but have now largely dropped their investments in these forms of renewable power in favor of biofuel development. Why? According to Jim Thomas from the technology watchdog ETC Group: 'It's because it's something they're already used to dealing with – a liquid fuel they can put in cars – and they have the existing infrastructure to handle and refine it.'[5]

This made liquid biofuels a more profitable prospect for oil companies than other forms of 'renewable' energy, allowing them to talk about developing alternative energy sources with less risk to their bottom line.

At the same time, powerful agricultural interests – from US corn growers to Brazilian sugar-cane magnates – saw the potential for a whole new fuel-based market for their products. The seeds of the biofuel boom had been sown.

Pressure from these industries, combined with other local factors (such as calls for energy security and 'rural development' in the US, and greenhouse-gas reduction targets in the EU), spurred governments around the world into supporting biofuel production and consumption. Now, 62 countries (including the 27 members of the EU) have a mandated 'blending' target – a pledge to mix a certain percentage of biofuel into their transport fuel supply, typically between 5 per cent and 20 per cent. These targets are often backed up by subsidies and tax breaks.[6]

In 2011, biofuels were approved for use as a fuel for commercial flights. KLM Royal Dutch Airlines began mixing 25 per cent biodiesel from waste cooking oil into its aviation fuel for certain European flights; as of 2014, the airline also carries out some flights between Amsterdam and New York with this fuel mix.[7] It has set a target of powering one per cent of its flights with biofuel by the end of 2015. Other airlines, such as Boeing and British Airways, have stated their support for similar targets. The aviation industry body IATA (International Air Transport Association) has stated that the industry is aiming to halve its greenhouse-gas emissions by 2050, with 'alternative fuels' as one of the most important routes to achieving this.[8]

By guaranteeing ever-increasing demand for biofuels, these policies from governments and airlines have flung open the farm gates to large-scale energy crop development. Production of bioethanol and biodiesel grew rapidly through the 2000s, particularly in the US, EU and Brazil; however, the rate of growth has now slowed, for reasons we'll explore in a moment.

Global production of ethanol from food crops reached 87 billion liters in 2013, while biodiesel production from food crops and livestock stood at 24 billion liters. Production of these fuels is highly concentrated – between them, the US and Brazil are responsible for 85 per cent of global ethanol fuel, while the EU, Brazil and the US together produce 70 per cent of global biodiesel.

The great majority of this is blended into these countries' transport fuel supply – around five per cent of the EU's transport fuel is now biodiesel, and 10 per cent of US vehicle fuel is ethanol.

A number of experimental plants producing 'second generation' cellulosic ethanol from stems, leaves and husks have opened in the past few years, mainly in the US, Brazil and Italy. If we add up the amounts of fuel that these plants claim they will produce, it comes to just under 0.3 billion liters per year. There are a number of other projects proposed for the next few years, especially in China and Denmark; if they all open and operate as planned, they will add another 0.9 billion liters, bringing total cellulosic ethanol production to 1.2 billion liters.

With the exception of Italy, all of these plants are planning to use agricultural waste, rather than purpose-grown crops, as their fuel source – at least for now.

Where does it all come from?

The biggest producers of biofuel (the US and Brazil) are mainly using home-grown crops to make it. Around 40 per cent of the US corn crop now feeds cars rather than people. A third of the fuel in Brazilian cars started life as sugar cane.

This doesn't quite tell the full story, though. There is a growing trend for agricultural companies to secure large areas of land in poorer countries, to produce crops that can be transformed into biofuels for richer consumers elsewhere. For example, at least four million hectares of land in Africa has been set aside for oily crops like jatropha and palm,[9] specifically for export to the UK, Germany, Italy, France and the US. A Chinese company is negotiating for up to a million hectares in the Democratic Republic of Congo, for growing palm oil for biodiesel.[10] According to Oxfam, around 227 million hectares of land has been bought or leased in developing countries by companies, foreign governments or other private interests since 2001, most of it in Africa.[11] Somewhere between a third and two-thirds of these land deals are believed to be for biofuel production.[12]

The world's top burner of wood in power stations is the United States, producing around a fifth of the world's biomass electricity all by itself. This is largely due to a rapid expansion in biomass electricity that took place in the US following the oil price rises of the 1970s. Legislation passed in 1978 created incentives for power stations to burn wood and agricultural waste.[13] As a result, the amount of US electricity generated from biomass rose 100-fold in a decade, from less than half a TWh in 1980 to almost 50 TWh in 1990. These power stations, which burn wood, corn ethanol and other agricultural products and wastes, now provide half of all of US 'renewable' electricity.

For other major producers of biomass electricity, growth has been more recent. Both Germany and the UK quadrupled the amount they generated between 2000 and 2012; China's output grew from 2 to 44 TWh between 2009 and 2012. The world now produces more than 20 times as much electricity from biomass as it did in 1980.

Costs, risks and drawbacks

While fuel crops are 'renewable' in theory – you grow them, you burn them, you grow some more – in practice they have one very real and very important limitation. They need land, and lots of it.

The possible exception to this is fuels made from

agricultural waste and used cooking oil. We'll look at these in more detail in Chapter 7, so for now let's focus on fuel made from purpose-grown energy crops (which makes up the great majority of biofuels, as we've seen).

The 'first generation' biofuels made from food crops need good, farm-quality soils. The problem is, the world's productive land tends to have things already on it: farms, or forests, or swamps and deltas rich in biodiversity. Growing fuel crops usually means either switching farmland over from feeding people to feeding cars, or destroying woodlands, rainforests and marshlands that not only act as plant and animal habitats but also as important stores of carbon and, often, as a source of food and fuel for local people.

Current ethanol and biodiesel fuel production uses an estimated 45 million hectares of farmland,* enough to feed almost half a billion people.[14] At least 17 million hectares of land have been seized from local communities in Africa, Asia, Latin America and Eastern Europe since 2002, and turned over to biofuel production.[15] This has forced countless farmers off their land and into poverty and hunger. Elsewhere, the biofuel industry provides support for slave-labor-style sugar-cane plantations in Brazil[16] and deforestation fuelled by soya plantations in Argentina and palm oil in Indonesia.

Switching any farmland over to fuel production means there's less land available for food; this pushes up the price of food and creates more demand for farmland elsewhere, which has all kinds of knock-on effects. The use of biofuels is thought to have played a significant role in the recent spikes in global food prices (along with food speculation, more meat-heavy diets and the impacts of climate change).[17] The US, as the world's biggest exporter of corn, is seen as a particular culprit here.

* According to UNEP, 35.7 million hectares were used to produce 81 billion liters of biofuel in 2008. If we assume that the land requirements of first-generation crops were roughly the same in 2013 as in 2008, 113 billion liters of biofuel would need 45 million hectares.

There's also a big question mark over the climate benefits of using fuel crops. Once the greenhouse gases from the fertilizers, farm energy, refining and transport of the fuels are taken into account, ethanol and biodiesel are far from zero carbon. Then we need to add in the carbon dioxide from 'Indirect Land Use Change' (ILUC). Switching a piece of food-growing farmland over to biofuels means that someone elsewhere will need to grow some extra food to compensate – and they may well be felling a forest or carving up peatland to do it. This leads to sizeable 'indirect' carbon emissions – see Chapter 8 for more details.

Can't see the trees for the wood

As we've seen, woods, grasses and other non-food plant materials are already commonly used as a heating fuel, and are increasingly being used to generate electricity or to manufacture 'cellulosic' liquid fuels. The use of these materials can create many of the same problems as liquid fuels, especially when driven by the needs of a profit-making industry.

While in theory, fast-growing trees and grasses could be grown on less fertile land unsuitable for food crops, in practice the companies involved prefer to use the most productive land available in order to maximize their profits. This means creating plantations on land that is currently covered by forests or used for agriculture.

Drax the Destroyer

Drax power station in North Yorkshire burns more coal than any other power plant in Britain – and now more wood than any other plant in the world. Drax has converted one of its six units to run on wood pellets, and so now burns through pellets made from five million tonnes of fresh 'green' wood per year. This is the equivalent of half of Britain's annual wood production. Drax's wood fuel is currently imported from the US and Canada, from regions infamous for the mass destruction of intact forests for pellet production. Drax aims to at least triple its wood intake, up to 1.5 times the level of British wood production.[18]

Meanwhile, using existing forests and woodland as a source of fuel can lead to serious deforestation if trees are taken faster than they can regrow. This problem can be caused by large numbers of people collecting firewood for home use or, increasingly, by companies gathering millions of tonnes to pour into power stations.

The International Energy Agency (IEA) estimates that – based on current global trends – four to six times more wood could be used for electricity (and heat) in 2035 than in 2009, with around 20 per cent of it being burnt in the EU. This is far too much wood to source from within Europe – it would need an area of plantations at least the size of Greece. Where would this wood come from? Brazil, Uruguay and other South American countries are already major exporters, and biomass investors are now acquiring land and plantations in other countries such as Ghana, Guyana, Congo, Uganda, Cameroon, Madagascar, Mozambique and Tanzania. The sheer scale of the EU's proposed biomass expansion makes it a similar but much bigger development problem than liquid biofuels – particularly in terms of climate change, land grabs and hunger. And this is just for the EU, even before we consider the projected growth in wood use elsewhere in the world.

The EU currently rates wood as a 'zero carbon' fuel, but this is a highly dubious claim to make, especially at this scale. Even if a new tree is planted for every tree that's burned, it can take decades for the new tree to absorb as much carbon as was lost from the old one. No firm standards yet exist to prevent wood fuel being imported from ex-rainforest plantations in South America or Africa, or from forests that are being cut to the ground in Canada and the US. Even if the trees are grown on existing plantations, there's still the 'ILUC' problem: turning plantations over to making fuel pushes demand for other wood and paper products elsewhere, which can lead to the same kinds of deforestation and land-grabbing as for liquid fuels.

Also, a major reason why so many energy companies have seized on wood fuel is because they can mix it in with coal in their already-existing coal power stations. This allows these stations to be classified as less polluting and thus able to continue operating, while other stations burning 100-per-cent coal are being shut down in line with EU climate regulations. This means that wood, rather than replacing coal, is in many cases allowing coal power stations to remain open.

Burning wood also creates air pollution, particularly volatile organic compounds and particulates. This can impact on the health of people living near wood-fired power stations, as well as people burning wood in open fires in their own homes.

Cellulosic fuels are in their early stages, and whether

Algal fuels

You may have heard about these 'wonder fuels' – sometimes called 'third generation' biofuels – that are currently in development. The idea is to breed (or genetically modify) algae to produce a natural oil, which could then be used to make biodiesel.

Some research is focused on macro-algae (seaweed), while others are working on micro-algae (pond scum). The plan is for these to be grown in the sea or in tanks, without the need for agricultural land.

Sounds neat, huh? Except, of course, that the researchers don't expect these fuels to be available on a useful scale for at least a decade, even if all goes well. Plus, once we scratch beneath the surface we find a whole host of potential problems that would need to be solved.

The huge seaweed farms proposed for the ocean could cause the same sort of problems – reduced biodiversity, unpredictable knock-on ecosystem effects and fertilizer leakage – as land-based monocultures. Tiny oil-producing algae in tanks still need large amounts of space, water, sunlight – and, crucially, nutrients. No-one is clear where these nutrients will come from; the algae will be competing with food crops here, especially for phosphates. The algae are expected to be genetically modified to churn out as much oil as possible – so what happens if (or when) they get loose into streams, rivers and reservoirs?

they go down the same route will depend on who develops them and for what purpose. At the moment, they are mainly being manufactured from agricultural waste, but there is obviously a limited amount of this available (see Chapter 7). If cellulosic fuels start to expand, the companies involved will soon run out of 'waste' to process – and then where will they turn?

Biofuels and biomass may have some role to play in a sustainable future, but not in the way that they are currently being exploited by agribusiness and the energy industries. We'll come back to them again in Chapter 10.

1 nin.tl/biofuelsdamage **2** Calculated from figures from Richard Murphy, Jeremy Woods, Mairi Black, Marcelle McManus, 'Global developments in the competition for land from biofuels', *Food Policy*, 36 (Suppl 1), S52-S61. **3** See for example nin.tl/1x5YjVM **4** nin.tl/biofuelwatch **5** nin.tl/biofueldamage **6** nin.tl/biofuelmandates and globalrfa.org/biofuels-map **7** nin.tl/klmbiofuelflight **8** nin.tl/1x5YJv8 **9** nin.tl/buyingfarmlandabroad **10** nin.tl/chinalandgrab **11** nin.tl/landandpower **12** nin.tl/africabiofuels **13** nin.tl/1balGn5 **14** Using a recent estimate that 10.1 people could be fed per hectare of cropland: nin.tl/1BkVBXt **15** nin.tl/grainlandgrabs **16** nin.tl/1Dni0dC **17** nin.tl/mealspergallon **18** nin.tl/globalforest-drax

7 Energy from 'waste'

> *'Everything must go somewhere. There is no "waste" in nature and there is no "away" to which things can be thrown.'*
> – Barry Commoner, *The Closed Circle*, 1971

> *'I come from garbage.'*
> – Shia LaBeouf

Does waste *really* count as a source of renewable energy?

This is a bit of a thorny question. Many countries, regions and states – including the EU – define energy from 'biogenic' waste as renewable. Biogenic waste is waste from living matter, such as discarded food, wood, paper and other natural materials.

Why is energy derived from these materials classed as renewable? Well, new plant material can, in theory, be grown to replace the stuff that was used for energy. Unlike fossil fuels, waste plant matter can be replenished within human time frames and so is counted as renewable.

Things aren't quite that simple, though. There are limits to how much plant material we can sustainably grow, because there's only a certain amount of productive land in the world – and a lot of competition for that land.

For example, one potential source of energy – as we'll see below – is wasted food. This can be converted into fertilizer and useful methane fuel. But that doesn't mean that a good way to solve our energy problems would be to throw away more food! That would require lots more land for growing the food in the first place, creating all

Organic waste? Sounds expensive...

When we talk about waste, the word 'organic' simply means 'made from stuff that used to be alive'. It's not the same meaning as the word organic in 'organic food', which refers to food that's been produced without using chemical pesticides or fertilizers.

kinds of problems all over the world (see Chapter 6).

Plus, of course, it takes energy to produce food, paper, wood, cardboard, cotton, and all the other natural products that could be counted as 'renewable' if we turned them into fuel at the end of their lives. The energy we save by avoiding (or failing that, recycling) a tonne of this waste is greater than the energy we would gain by turning a tonne of waste into fuel. Reducing waste is always the best option.

So let's be absolutely clear. Avoiding waste, or recycling it into new products, will reduce demand for energy and land. This will do much more to make a 100-per-cent renewable world possible than continuing to chuck lots of stuff away but then turning some of that waste into fuel.

However, there's a certain amount of naturally produced stuff that we are, unavoidably, going to throw away. Broken timber from refurbished houses. Teabags, coffee grounds and the green bits off the top of carrots. Certain agricultural wastes that can't be ploughed straight back in the soil. The cauliflower that your four-

The wrong energy from the wrong waste

Burning general domestic waste to generate heat or electricity is *not* a form of renewable energy. One major reason is that household waste nearly always contains a lot of materials containing plastic. Because plastics are made from oil and gas, burning waste synthetic materials (whether it's plastic packaging or nylon clothing) for energy has the same effect as burning fossil fuels – it releases fossil carbon into the atmosphere.

This is one reason why the monstrous chuck-it-all-in waste incinerators that some countries like to build as an alternative to landfill should not been seen as a source of 'renewable' energy. They burn a lot of plastic, as well as material that really should have been recycled, composted, or never made in the first place. They also stand accused of creating local air pollution and discouraging recycling. They may be slightly more energy-efficient than landfill sites, but that doesn't really make them a good thing.

year-old child absolutely loved last week but inexplicably won't eat today and has dropped all over the dog. We are unlikely to totally eradicate all these kinds of wastes any time soon, so if there are ways to capture useful energy from them then we might want to give that a go. Some of the more promising ways to do this are explained in this chapter.

Types of 'renewable' energy from waste

Anaerobic digestion

This is a particularly useful technique for domestic food waste, sewage sludge, and agricultural wastes that can't be directly composted or re-used on the farm. 'Anaerobic' means 'without oxygen', and so an anaerobic digester is a device where organic waste is broken down without exposure to the air. This favors particular bacteria that helpfully break down the waste and release 'biogas'; this is a flammable gas mainly made up of methane, with a bit of carbon dioxide and some other things mixed in. It can then be burned directly as a local heating fuel, or refined into a purer 'biomethane' that can be piped into natural gas grids or used as a vehicle fuel.

Figure 7.1: Anaerobic digesters in Tel Aviv, Israel

The process also produces solid and liquid 'digestate'. The solid digestate can be used as a soil improver, while the liquid makes an excellent fertilizer. This is the beauty of anaerobic digestion – it transforms rather unpleasant waste into several useful outputs.

Waste cooking oil

Will we still have fast food in the future? Whether the thought of deep-fried food makes your stomach rumble or turn, the fact is that there's a lot of it out there at the moment, which means a lot of used oil being thrown away. Add in all the waste cooking oil and fat from homes and food-processing plants and it adds up to quite a greasy mountain.

Over the last 20 years, there's been growing interest in turning this waste into fuel. The US currently disposes of up to 11 billion liters of waste fat each year,[1] and initiatives and companies have sprung up all over the country to collect and reprocess this waste into biodiesel (using the same process as used to make biodiesel from fresh vegetable oils). Similar projects have been launched all over the world.

It's worth noting that fuels made from natural fats are

Figure 7.2: The 'Big Lemon', a bus run on 100-per-cent used cooking oil, Brighton, UK

often referred to simply as 'biodiesel', whether they're processed from recycled oil or made from new crops. This can be confusing, so watch out for it.

Gas from water treatment and landfill

Waste water treatment plants and landfill sites both emit methane (and other gases), from the rotting organic matter they contain. In many countries, these sites have been adapted to collect this 'biogas' mixture, which can then be used as an onsite fuel or refined into purer 'biomethane' for other purposes. It's not as efficient as collecting the organic waste separately and putting it in an anaerobic digester, but it's better than just letting the methane from landfill and sewage escape into the atmosphere, where it would act as a powerful greenhouse gas.

Direct heat from waste

Put simply, some materials such as wood waste from building sites or forestry could simply be burned to provide heat for homes and workplaces. However, this should only happen to materials that definitely can't be re-used in some way, and don't have an important role as part of forest ecosystems.

History and current use

Humanity has been turning waste into fuel for thousands of years. Animal dung has a long, proud and noxious history as a fuel for heating and cooking, and is still used by many people around the world (though usually from necessity rather than choice).

However, before the Industrial Revolution – and particularly before the rise of consumerism – human society wasn't really producing enough waste for it to be considered a fuel source (unless we include feeding food scraps to livestock).

Other than direct burning, the energy capture techniques described in this chapter have only been seriously developed over the last few decades. Although the first anaerobic digester was built at a leper colony in India in 1859,[2] the technique didn't really become widespread until the late 1900s. Anaerobic digesters have particularly taken off in China and India, where around 30 million households now make cooking and heating gas with their own domestic devices. Some of these devices also generate electricity for lighting.

Some local authorities, especially in Europe, are now experimenting with collecting household food waste separately from other rubbish, and putting it into anaerobic digesters. In 2010, the first anaerobic sewage digester opened at Didcot in the UK, and provides enough gas to heat 200 homes.

Costs, risks and drawbacks

Using waste as fuel is a relatively cheap way to generate energy, especially in countries with high taxes and fees for waste disposal. The main risks are tied into the fact that it is always better to avoid waste in the first place rather than use it as a fuel, because reducing waste saves more energy overall. If we become too reliant on energy from waste, we could end up creating incentives to be more, rather than less, wasteful. Anaerobic digesters need to be fed with waste for a number of years in order to pay back their upfront costs; once they've been built then their owners and employees will have a strong incentive to keep the waste flowing in.

Using 'unavoidable' waste as a source of energy makes sense, but avoidable waste needs to be, well, *avoided.* If we build an energy industry based on avoidable waste then that's going to cost us more energy than it brings us in the long run. So this is something that needs to be managed very carefully and used in a limited way.

1 ybiofuels.org/images/56July.pdf **2** P-J Meynell, *Methane: Planning a Digester*, Schocken Books, New York, 1976, p 3.

8 Renewables versus fossil fuels

Our civilization runs by burning the remains of humble creatures who inhabited the Earth hundreds of millions of years before the first humans came on the scene. Like some ghastly cannibal cult, we subsist on the dead bodies of our ancestors and distant relatives.
– Carl Sagan

In this chapter, we'll take a look at how renewable sources stack up next to fossil fuels. We'll briefly compare their environmental impacts, the amount of energy they currently generate, and how much they might be able to generate in the future.

Energy payback of renewables

The tired old 'it takes more energy to build a wind turbine/solar panel than it'll ever generate' myth gets dragged out a lot. In reality, modern renewable generation sources will pay back their energy costs many times over, so long as they're well sited and properly installed – see Figure 8.1.

Figure 8.1 Energy payback for different electricity sources

	Time taken to generate the energy used to build and run it	How many times it will pay that energy back over its lifetime
Solar panel[1]	10 to 30 months	10 to 35 times
Wind turbine[2]	3 to 6 months	At least 20 times
CSP plant[3]	12 to 13 months	10 to 30 times
Small hydro[4]	5 to 16 months	75 to 225 times
Geothermal plant[5]	12 to 48 months	10 to 30 times

Greenhouse-gas emissions

We can hope that we'll eventually manufacture all of our generation technology using 100-per-cent renewable energy, but at the moment we are still burning some

Figure 8.2 Greenhouse gas emissions per KWh, for fossil fuels and renewables[6]

	Total emissions including construction, fuel use and knock-on effects (gCO_2e per KWh)*
Rooftop PV	26 - 60
PV solar farm	18 - 180
Onshore wind	7 - 56
Offshore wind	8 - 35
CSP	9 - 63
Large hydro[7]	1 - 2,200
Small hydro	1 - 20
Geothermal electricity	6 - 79
Wave and tidal	6 - 28
Biomass electricity	130 - 890
Electricity from mixed waste incineration	390 - 530
Electricity from waste gas	90
Electricity from coal	740 - 910
Electricity from gas	350 - 490
Electricity from oil	710 - 2,100
Electricity from nuclear	4 - 100
Heat from air, ground and water source heat pumps	35 - 45 (if powered by renewable electricity)
Solar water heating[8]	5 - 6
Heat from coal	330 - 430
Heat from gas	220 - 250
Heat from oil	300 - 330
Fossil oil for transport	240 - 270
Bioethanol for transport[9]	140 - 290
Biodiesel for transport	340 - 450

* gCO_2e means 'grams of CO_2 equivalent'. This unit is commonly used as a measure of global warming impact.

fossil fuels in the process. However, spread across their lifetimes, the greenhouse-gas emissions from renewable generators (apart from dams and biofuels) are very low, as shown in Figure 8.2.

Impacts on wildlife

We've already seen that wind turbines can have an impact on birds. Some CSP plants have also been linked with the deaths of birds flying through the concentrated solar rays. Figure 8.3 shows how these impacts compare with damage to birdlife from other energy sources, including the effects of collisions, fossil-fuel extraction, uranium mining and climate change.

Large dams and tidal barrages typically have a serious impact on aquatic and semi-aquatic life. The impacts are hard to quantify as they vary from scheme to scheme, but can include wiping out entire habitats. Smaller run-of-the-river hydro projects can also have impacts, but careful design and siting can reduce these greatly.

Land use

Figure 8.4 compares the land space required for different electricity sources.

Water use

A specific concern for desert-based solar is the water required to clean or cool the system. For CSP, this can be as much as 3,500 liters per MWh, compared with 2,000 l/MWh for new coal power plants and 1,000 l/MWh for the most efficient gas plants. However, 'dry cooling' technology deployed in Spanish CSP plants has successfully reduced this figure to 35-40 liters per MWh.[13]

Mining and materials

Figure 8.5 compares the raw materials required per GWh of electricity generated from different sources over their lifetimes.

Figure 8.3: Bird deaths from different forms of electricity generation[10]

	Average bird deaths per GWh generated
Solar PV	*negligible*
Wind (onshore and offshore)	0.3
CSP[11]	0.6
Nuclear plant	0.6
Fossil-fuelled power station	9.4

Figure 8.4: Land area required for different forms of electricity generation[12]

	Square meters used per GWh generated per year
Solar PV	7,500
Wind (onshore)	7,300
CSP	3,200
Nuclear plant	1,200
Coal power station (with strip mining)	5,700
Geothermal plant	160-290
Large hydro dam	250,000

We'll look at what these figures mean for scaling up renewable energy to the whole world in Chapter 10.

Financial costs

No-one has yet found a way to charge us for the wind or the sun (thank goodness), so once a turbine or solar panel is up it's incredibly cheap to run because there's no fuel to buy. However, to calculate the full price of renewables we need to include the costs of building, maintaining and eventually dismantling the equipment.

The official industry method for comparing the

Figure 8.5: Average tonnes of fuel (coal or gas) and building materials (iron, cement, aluminum and copper) required per GWh of electricity generated[14]

	Tonnes fuel	Tonnes building material	Total
Coal power station	490	1.0	491.0
Gas power station	160	0.4	160.4
Solar PV	0	2.7	2.7
CSP	0	6.8	6.8
Wind (Onshore and Offshore)	0	6.1	6.1
Hydro	0	3.8	3.8

cost of energy sources is the 'Levelized Cost of Electricity' (LCOE), which uses installation, running and decommissioning costs to give an average price per KWh generated over the life of the turbine, power station, solar panel or whatever. These figures differ from country to country.

In the US, onshore wind is now the cheapest form of energy generation, with an LCOE ranging from 3.7-8.1 cents per KWh. Offshore wind ranges from 6.0-18.0 cents per KWh. These figures compare with 6.1-8.7 cents for new gas plants, 6.6-15.1 cents for coal plants, 9.2-13.2 cents for nuclear, and 6.0-26.5 cents for solar.[15,16]

A 2014 Ecofys study reported that new onshore wind power in the EU cost roughly the same as new coal power and slightly less than new gas power.[17] However, the same report also considered what the price of these forms of energy would be if their environmental, climate and health costs were also taken into account.[18] By this measure, onshore wind was the cheapest source, with offshore wind and solar electricity not far behind. Gas-powered electricity was 50 per cent more expensive than onshore wind per KWh when these wider impacts were included; coal was more than double the cost of onshore wind, and almost twice the price of offshore wind.

Figure 8.6: Quantities of global renewable generation (2013)

	Amount installed globally (2013)	Energy generated per year (TWh)
Electricity		
Solar PV[19]	139 GW	160
Wind[20]	318 GW (approx 250,000 turbines)	620
CSP plant[21]	3.4 GW	9
Geothermal plant[22]	12 GW	76
Tidal barrages	0.5 GW	1.2
Large hydro[23]	1,200 GW	3,515
Small hydro[24]	90 GW	265
Biomass power plants[25]	A few thousand stations	405
Heat		
Biomass and waste in Global South[26]	Used by over 2 billion people	9,000
Biomass and waste in 'modern' stoves	Many thousands of buildings and industrial processes	3,700
Solar water heaters	Approx. 50 million	240
Solar cookers[27]	Approx. 1.5 million	1.6
Passive solar	*Total unknown*	*Total unknown*
Heat pumps[28]	Approx. 150 million	2,800
Geothermal heat	Concentrated in a few countries	90
Transport fuel		
Bioethanol	Mostly in US and Brazil	520
Biodiesel	Mostly in the EU	260
Waste cooking oil	Mostly in the EU, China, US	25

Renewable sources also have the benefit of a very stable price; once you've built a wind farm or solar plant you have a very good idea of what it'll cost you to run because the price of the wind and sun (nothing) doesn't change. Coal and gas prices, on the other hand, can be very volatile and change rapidly, making the running costs of fossil-fuel plants harder to predict.

Adding it up

Figure 8.6 shows how much energy we currently derive from all the renewable sources laid out in the previous seven chapters.

So just over 5,000 TWh of 'renewable' electricity was generated in 2013, of which 70 per cent was from big dams and barrages, and 8 per cent from biomass power stations. Electricity from less problematic sources (wind, solar, geothermal, small hydro) stood at 1,130 TWh. This compares with global electricity use of 18,900 TWh in 2013.[29]

Currently, 80 per cent of 'renewable' heat comes from burning wood and other organic materials. It's hard to say how much of this is genuinely sustainable. The 12,700 TWh of biomass heat and 3,130 TWh of other renewable heat compares with global heat use of 41,800 TWh in 2013.[30]

The 780 TWh of liquid biofuels and 25 TWh of waste cooking oil fuel compare with a global demand for liquid fuels of around 49,000 TWh per year.

So solar power is currently providing just under one per cent of global electricity use, while solar water heating and solar cooking are providing just over half a per cent of global heat demand. Together, these forms of solar power are providing 0.4 per cent of global energy demand. Solar energy may be growing fast, but there's clearly still a long way to go.

Wind is currently providing 3.3 per cent of global electricity demand, and 0.6 per cent of global energy. However, this figure is much higher in certain places.

Figure 8.7: Potential annual energy production from renewable energy, using existing technology only

	Potential annual energy production (TWh)	Source/Notes
Electricity		
Solar PV on roofs[31]	15,000	David MacKay
Solar PV in 'farms'	370,000 to 4,100,000	IPCC[32]
Wind	53,000	European Wind Energy Association
CSP	70,000 to 220,000	IPCC[33]
Geothermal plant	33,000 to 308,000	IPCC
Wave power[34]	4,300 to 5,500	IPCC (early estimate)
Tidal power	800	Paul Lynn (early estimate)
Large hydro	14,500	IPCC
Small hydro	610	UN[35]
Biomass electricity	*See Chapter 10*	
Heat		
Biomass	*See Chapter 10*	
Waste gas	3,700 to 4,700	Using 'unavoidable' waste only (see below)
Solar water heaters[36]	8,000	David MacKay
Solar cookers	330	300 million households (Solar Cookers International)
Heat pumps	13,500	If 9 billion people used an average of 1,500 KWh/year (see below)
Geothermal heat	3,000 to 88,000	IPCC
Transport fuel		
Purpose-grown fuel crops	*See Chapter 10*	
Waste cooking oil	120 to 150	If all countries use similar amounts of cooking oil per person as the UK

Wind power made up 32 per cent of all new electricity generation installed in the EU in 2013, and is now meeting 8 per cent of EU electricity demand.

How much could existing technology provide?

Figure 8.7 gives some estimates for how much energy could be produced by the various renewable technologies if they were rolled out worldwide.

For comparison, we currently burn through 155,000 TWh of energy per year, mostly from fossil fuels. A brief glance at this table might make you think that we could easily replace that total with renewables; however, in reality things aren't quite that simple, as we'll see in Chapters 9 and 10!

Meanwhile, some of the numbers in the table need a bit more explanation.

Solar power

You'll notice a *really* big maximum figure for PV – over four million TWh, more than 25 times current global energy use. To reach this maximum, we'd need to cover two per cent of the earth's land surface with solar generators – around 300 million hectares, about the size of India and equivalent to a quarter of all the world's farmland. That's just for PV, before we even think about CSP plants too.

This seems rather extreme, to say the least. Covering quite that much land with solar farms would clearly have all kinds of negative knock-on effects for food production, landscapes and ecosystems. We'll look again at how much of this solar power we could realistically develop in Chapter 10.

Wind

To get this figure, the EWEA study looks at all the spots on Earth that are windy enough for turbines to work, then assumes that we'll build wind generation in less than 10 per cent of these places.[37] It also assumes we'll use existing technology, and that wind turbines don't improve in the future (although they probably will).

Geothermal

The wide range reflects the fact that we haven't yet scoped out all the possible spots that could have good geothermal resources, and that we also don't know how good our enhanced geothermal technology is going to be. However, even the lower numbers suggest that geothermal could play a serious role in our renewably powered future.

Heat pumps

The global energy potential depends on a range of local factors – not just the amount of heat available in the air, ground and water near people's homes, but also the demand for heat. Regions that stay warm all year round will have little use for ground- or air-source heating.

Researchers at the Centre for Alternative Technology in Wales suggest that widespread use of heat pumps in the UK could provide an average of 1,500 KWh per person per year by 2030. This seems like a fairly conservative figure: the International Energy Agency suggests that a single domestic heat pump can produce 15,000 KWh per year, while a commercial-sized one can produce 100,000 KWh per year. Sweden already derives over 3,000 KWh of heat per person from heat pumps, thanks to a combination of millions of installed pumps and high energy demand during their very cold winters.[38]

It therefore seems fair to use 1,500 KWh/person/year as a cautious estimate of the total energy that the world might harness from heat pumps, allowing for the fact that people in some places might be able to draw on a much higher amount than this, while in other places it wouldn't really be suitable at all.

Bioenergy

It's incredibly tough to put a figure on how much bioenergy we could sustainably use. It's already a huge source of energy, but – as we saw in Chapter 6 – very little of it is being produced or used in a sustainable way. If we want to use bioenergy in a cleaner, fairer future, we'll need to obtain and harness it in a very different way from today.

We'll look at this again in Chapter 10, and, instead of trying to work out a 'sustainable maximum', we'll ask: what's the minimum amount of bioenergy that we need, and is it possible to produce it sustainably?

Waste gas

The Zero Carbon Britain report from the Centre for Alternative Technology in Wales includes a careful assessment of the amount of energy that could be generated from 'unavoidable' waste in a UK run on entirely renewable energy. Their model assumes that waste is greatly reduced and that 83 per cent of food is produced within the country (an increase from 58 per cent today). In this scenario, 5,200 KWh of biomethane could be generated per person from the anaerobic digestion of food, agricultural and sewage waste.

If we scaled this up for the world – and made sure everyone on the planet had enough to eat – that would give us 3,700 TWh of biogas today, or 4,700 TWh in 2050 (if the world's population grows to 9 billion, as predicted by the UN).

1 nin.tl/19nvWYk **2** Ida Kubiszewskia, J Cutler Cleveland, Peter K Endresc, *Renewable Energy*, Vol 35, Issue 1, Jan 2010, pp 218-225, Meta-analysis of net energy return for wind power systems. **3** nin.tl/1EmAfxv **4** nin.tl/1BHze3i **5** nin.tl/1wLARfR **6** nin.tl/IPCCmitigation **7** Small hydro schemes in rocky environments produce almost no methane, while projects that flood large areas of tropical vegetation can produce between 1,000 and 2,000 gCO_2e per KWh. Source: nin.tl/hydropoweremissions **8** nin.tl/1x5ZQeq **9** The figures for bioenergy in this table include the official EU estimates for ILUC (see Chapter 6). Some critics claim that these figures have been underestimated. Source: nin.tl/1AGtfqw **10** nin.tl/fossilfuelcost **11** Based on figures from the Ivanpah 'power tower' CSP plant in California, which has come under particular criticism for this. Ivanpah is now trialling different bird deterrents to try to reduce the problem further. **12** Ronald Dipippo, *Geothermal Power Plants: Principles, Applications, Case Studies and Environmental Impact*, Elsevier, 2012. **13** nin.tl/CSPwater **14** nin.tl/1xp1Rgn **15** nin.

tl/1GH9FPA **16** cleantechnica.com/wind-energy-facts **17** nin.tl/EUenergysub-sidies **18** It is, of course, impossible to put a financial value on things like deaths from air pollution and the threats that climate change poses to our entire civilization and way of life. The numbers in the Ecofys report for the financial damage caused by fossil fuels should be treated with caution, as they only measure things like the costs of extra healthcare or the financial costs to agricultural production from climate change. They shouldn't be seen as measuring the full costs of fossil fuels because you simply can't put a financial value on a human life or a species driven to extinction. While these numbers are useful for making a quick point about the fact that fossil fuels have serious hidden costs that make them more expensive than renewables in cold financial terms, I wouldn't recommend using them more widely than this. **19** IEA's Photovoltaic Power Systems Programme. **20** Global Wind Energy Council. **21** Annual generation from my own calculations, based on the average production from Spain's CSP plants. **22** nin.tl/1GH9XWG **23** Calculated from nin.tl/smallhydroworld and ren21.net **24** Calculated fromnin.tl/smallhydroworld and ren21.net **25** ren21.net **26** Calculated from ren21.net data **27** Based on an average of 3 KWh per day, from a study in Argentina. **28** According to the European Heat Pumps Association, the 6.7 million pumps in Europe are currently producing around 121 TWh of heat, so that's an average of 18,000 KWh per pump per year. If we assume that 8 million extra heat pumps have been installed worldwide since 2008, then that's 153 million pumps x 18,000 KWh = 2,800 TWh per year. If we assume they're working at an average coefficient of 3, then they're using 900 TWh of electricity to produce that energy, so that's a net gain of 1,900 TWh of clean energy. The energy collected by heat pumps isn't officially counted by the International Energy Agency in their global energy totals. So it's an extra 2,800 TWh of heat energy that we're currently consuming on top of the official 104,000 TWh from other sources that the IEA does count. **29** IEA figures. **30** IEA figures, plus the estimated 2,800 TWh from heat pumps not counted by the IEA. **31** withouthotair.com **32** nin.tl/1AqJHvH **33** nin.tl/1AqJHvH **34** nin.tl/1CimElc **35** nin.tl/smallhydroworld **36** withouthotair.com **37** nin.tl/offshorewindpower **38** nin.tl/swedishheatpump

9 How much energy do we need?

'Don't mistake activity for achievement.'
– John Wooden

Chapter 8 gave us a rough idea of the amount of energy that different kinds of renewables could provide, using current technology.

But is this enough to power the world? Well, in order to decide that, the important question is: *how much energy do we need?*

This question doesn't get asked often enough, and, when it does get asked, there's a completely wrong answer that we often hear in response: 'Global energy use is growing, so if we track that on a graph we find we'll need a load more of it in 2035, or 2040, or 2050, and *that's* how much energy we need to produce.'

I'll explain in a moment why I think this is the wrong way to answer the question, but first it's worth noting that some people take these kinds of predictions very seriously. For example, the International Energy Agency (IEA) believes the world will use 37 per cent more energy in 2040 than in 2013.[1] BP thinks that energy demand will grow even faster than this, with energy use in 2035 41 per cent higher than in 2013.[2] Shell says that global energy use in 2050 could be three times higher than in 2000.[3]

These energy predictions from the IEA, BP and Shell would give us a global demand somewhere between 210,000 and 250,000 TWh/year by 2050. This is the kind of number that then gets thrown around in debates about future energy use. Renewable-energy fans like to point out that, although this is a lot of energy, there are studies that show we could reach these totals if we pushed for maximum deployment of solar, wind and geothermal energy (Figure 8.7).

Fossil-fuel (and nuclear) advocates, on the other hand, say this is far too much energy for renewables to handle; we are currently producing around 12,600 TWh from 'modern' renewable energy, and so would need to expand our renewable-energy production 20 times over in order to meet these targets. Oil companies like Exxon, Shell and BP have used these figures to suggest that renewable energy simply can't fill the gap and that, in 2050, we will still need fossil fuels to provide half of the world's energy. In fact, they are basing their entire business plans on this idea, even though it would mean runaway climate disaster for the whole planet.

There are two big reasons why we shouldn't use these industry projections as our 'target' for global energy use.

First, these projections assume that the world is going to carry on along the same path as today, with similar levels of global inequality, mass poverty, and the concentration of power and wealth in the hands of a minority.

The industry projections assume greater energy use by a growing middle class in countries like India, China, Brazil, Russia and South Africa, but that most of the world will still remain in poverty. If you dig into the reports, it becomes clear that the 35-40 per cent increase in energy use envisioned by the IEA and the fossil-fuel giants would – like today – mostly be consumed by a wealthy minority, and would not reach most of the planet's population.

I don't know about you, but that's not a future I want to be working towards. The reason I want to know whether a 100-per-cent renewable world is possible is so that I can go out and shout about it, to tell the naysayers and the cynics that we don't need fossil fuels. But I'm not interested in going out and saying 'hey everyone, we can power the world renewably, but only so long as most people stay poor'. Not only is that morally indefensible, it would also be doomed to failure. Right now, some of the most important work in building a clean-energy future is being done by frontline communities – people who are feeling the impacts of fossil-fuel extraction and climate change and are fighting back. This includes some of the poorest and most marginalized people in the world. Any clean-energy future that does not meet their needs is extremely unlikely to happen. We'll come back to this point in Chapter 11.

Second, the industry figures are answering the wrong question. They're estimates of how much energy the world might use in a particular future scenario. This isn't the same thing as the amount of energy we actually *need* – that's a different question entirely.

If we want to avoid the nightmarish future mapped out for us by the fossil-fuel companies, then we need to find a better answer to this question. How much energy do we really need – not just for unfair and destructive business as usual, but for a world where *everyone* has access to the energy required for a good life?

Living like a European

To answer this question, first we need to define it a bit more carefully. The amount of energy we need depends on the kind of lifestyle we think is important. If we think everyone *really needs* their own heated swimming pool and private jet, then we'll end up with a big number. If we think that it's fine for everyone to sleep under hedges, walk everywhere and eat nothing but wild blackberries then the amount of energy required will be small (also, we'll need to plant a lot more hedges and bramble bushes).

We probably want to find a happy medium between these extremes. One way to define this might be: we need everyone to have access to the energy they need for a good quality of life, while keeping any waste or excess to a minimum. This is important because – as shown in Chapters 1 to 7 – there's no such thing as completely 'clean' energy. Although renewables are much less dangerous than fossil fuels, they still have some negative impacts that we need to keep to a minimum.

When it comes to current levels of energy use, the European Union (EU) is sometimes seen as a sensible medium. Most people there have access to electricity, heating, food, travel, and a wide range of goods and services, but EU residents use significantly less energy per person than the US, Canada or Australia (see Figure 9.1).

Figure 9.1: Average per-person energy use in selected countries (International Energy Agency)

Country/region	Energy use per capita (KWh in 2013)
Canada	110,000
United States	83,500
Australia	58,500
European Union	38,500
China	24,500
The world's 80 poorest countries*	4,000
Nepal	1,000
Tanzania	800

There's another important factor we need to take into account: population. The world's population in 2015 stands at just over seven billion. The population is still growing, but that growth is slowing down; according to the UN, on current trends we will reach nine billion people by 2040, 10 billion by 2065 and then the population will level off at around 10.8 billion by the year 2100.

The UN would be the first to admit that these numbers are just rough estimates. No-one can really know what the population will be in 10 years' time, let alone 85 years, and there are all sorts of factors that could dramatically affect these figures.

I'm including them here, though, because we need to be able to say with confidence that a 100-per-cent renewable world is possible *even if the world's population continues to grow*. It's no good calculating that we've got enough renewable energy for seven billion people, if an extra two, three or four billion are about to show up.

* The 80 countries with the lowest GDP per capita according to the IMF. These countries account for 40 per cent of world population.

Let's say we want to have a world powered entirely by renewable energy by 2040. So long as we start the transition soon, that should give us a decent shot at avoiding the worst effects of climate change.*

If that's our aim, then we need to show that there'll be enough renewable energy available for the nine billion people that we're expecting to be here in 2040. On top of this, it would also be good to know: if the UN is right, and world population is going to eventually level off at around 11 billion people, would there be enough renewable energy for all of them, too?

So, back to the numbers in Figure 9.1. If we brought everyone in the world up to the average EU energy use of 38,500 KWh per year (apart from the minority who are using more than that, who would need to become more energy efficient and come down to this figure), how much energy would we need?

Before we do this calculation, we need to note that the numbers in this table are for *primary energy use*. This is the total energy consumed per person, but is not the same as the amount of useful energy actually delivered to homes, factories, farms and offices. The amount of useful energy supplied is called the *final energy demand*, and is a smaller number than the primary energy use. Why? Well, because a lot of the primary energy is lost along the way, especially when converting fossil-fuel energy into electricity. Up to two-thirds of the energy in coal, oil or gas can be lost as heat up the power-station chimneys. The same is true for wood-fired power stations. Further energy is lost when electricity is transferred along wires, or used to make hydrogen fuel.

Currently, final energy demand is around 30 per cent lower than primary energy use, thanks mainly to the energy

* To have a decent chance of avoiding runaway climate change, we can only afford to burn 20 per cent of known fossil fuels. If we carry on with current energy use patterns, we will pass that threshold by 2034. However, if we start cutting fossil-fuel use fast enough to get to zero by 2040, we should stay safely within that limit. nin.tl/carbonbudget2034

lost when fossil fuels are used to generate electricity. So 38,500 KWh of primary energy in the EU supplies around 27,000 KWh of final 'useful' energy. Getting 27,000 KWh of useful energy per year in a 100-per-cent renewable scenario would require just 33,500 KWh of primary energy per year, because we'd no longer be losing all that waste heat from burning fossil fuels.*

That means that, on a planet of nine billion people, we would require just over 300,000 TWh of energy per year if everyone used the same amount of energy as an EU resident. That's around three times current global energy use. If the world then grew to 11 billion, that number would reach 370,000 TWh per year.

Looking at the energy available from renewables (Figure 8.7), we can see that the total likely energy from the most benign and well-proven renewable technologies – wind, rooftop solar PV, rooftop solar heating, heat pumps, anaerobic waste digestion and small-scale hydro – comes to about 95,000 TWh per year.** To reach the EU average for the whole world, we'll need to find a further 205,000-275,000 TWh from other renewable sources.

CSP in deserts and geothermal power are only available in certain places and we don't yet know exactly how much could be realistically developed; enhanced geothermal also carries some risk of earthquakes and water contamination that we want to keep to a minimum. Placing solar PV on the ground rather than rooftops takes up land that is valuable for other uses, and using any kind of biomass or biofuel creates the same difficulty.

All of these energy sources could be useful in moderation, but we need to be cautious and sensitive in how we deploy them. Is it possible to get 205,000-275,000 TWh from these sources in a careful way that

---

* For twoenergyfutures.org, I calculated that the energy losses in a 100-per-cent renewable future would be 20 per cent, mainly from the conversion of electricity into hydrogen and other useful fuels.
** I'm not including wave or tidal power here because they're not yet fully proven on a large scale.

doesn't throw up a host of new problems and start encroaching on other natural limits? 205,000 TWh is about twice as much energy as we currently use globally. It *might* be possible, but I'd say it was extremely unlikely, especially when we take into account the raw materials needed to build all the equipment (see Chapter 10). A future that relies on pushing every possible renewable source to its limits feels like a very dangerous one, and one that isn't compatible with a respect for global justice and the rights of communities in areas where the energy is generated. Take a look at what's happened when companies and governments have charged ahead with renewable technologies like biofuels and large dams without paying proper heed to the consequences. Are these the kinds of examples we want to follow?

If this was our only option for a 100-per-cent renewable world, I'd be feeling pretty depressed right now. Luckily, though, there's another important avenue to explore, because we still haven't properly answered our original question: how much energy do we actually *need*?

Defining the good life

The thing is, that figure of 38,500 KWh of energy per person in Europe includes a huge amount of wasted energy. Across Europe today, thousands of millions of KWh of heat is leaking out of badly insulated houses. Millions of people are sitting alone in their cars in traffic jams, engines chugging. Factories are churning out bits of plastic tat, wrapped in more plastic, that will all be in a landfill site before the year is out. Is all this energy really adding anything important to our lives?

What if we did all the things that we've been told, time and time again, would save us energy while mostly making our life better? What if we really did have properly insulated and ventilated buildings that stayed warm in cold weather and cool in the heat?* What if we had decent

* Including the use of passive solar heating, as detailed in Chapter 1.

public transport, cycling facilities and car-sharing schemes that meant most people had no need to own a car? What if our food was organic and mostly locally produced, and if social success wasn't measured by how much stuff we bought but by how well we spent our time together? How much energy would we need then?

There are obviously lots of possible answers to this question, but a good starting place is the well-respected, cutting-edge research carried out at the Centre for Alternative Technology (CAT) in Machynlleth, Wales. Its 2013 report, *Zero Carbon Britain: Rethinking the Future*, lays out a blueprint for how the UK could be powered without the use of fossil fuels.

The Power Down section of this report explains how the quality of life in the UK could be maintained – or in many cases improved – while more than halving the amount of energy the country uses. This involves a major shift from car use onto public transport, serious retrofitting of homes and businesses to be more energy efficient, more local production of food and other goods, and everyone buying less consumer junk. It also requires the minority of people who currently fly a lot to cut down; instead, everyone flies the same amount as someone on an average UK income, which is one long-haul return flight every few years.

According to CAT's figures, the amount required for this less wasteful version of a British lifestyle is about 13,000 KWh per person per year.* In other words, it is perfectly possible to keep all the most important bits of a 'European' lifestyle – comfortable homes, good food, transport, decent healthcare, education, entertainment and so on – using less than half of the energy consumed by the average EU resident today.

What does this mean for our global energy targets?

---

* Actually, CAT put it a bit lower than this but I've increased it by 25 per cent to include the energy currently used to manufacture things overseas and import them to the UK. See twoenergyfutures.org for more information on the numbers behind this.

Taking into account the losses from converting primary energy into the right forms (electricity, hydrogen, liquid fuels) for final energy demand, we come to a grand total of 145,000 TWh for nine billion people, and 180,000 TWh for 11 billion.

We've already seen that we can get 95,000 TWh from wind, rooftop solar, heat pumps, waste digesters and small-scale hydro. That leaves us 50,000 TWh to find from CSP plants, PV farms, geothermal heat and electricity, wave, tidal and possibly some sustainably produced energy crops by 2040, rising to 85,000 TWh by 2100.

Seeing as the low-end estimates for CSP and geothermal electricity are 70,000 and 33,000 TWh respectively, this suddenly feels like a far more achievable challenge than when we were trying to match current EU energy use.

It seems clear that changing the way we use energy is going to be a vital part of reaching a 100-per-cent renewable world. But it's important to note three interesting consequences of saying that everyone in the world could have a 'good' life using just 13,000 KWh per year.

1. We'd be creating a much fairer world. Most people in the world currently use less than 13,000 KWh of energy per year. As can be seen in Figure 9.1, the residents of the 80 poorest countries use less than a third of this amount. For most of the world, we're talking about a major step upwards in access to energy.

2. However, this calculation only works if the currently energy-hungry countries adjust their demand downwards at the same time. If the richer minority stay on their current levels of per-capita energy use while everyone else comes up to 13,000 KWh, then global energy demand will rocket up past 200,000 TWh by 2040. In other words, the richest minority need to reduce their energy use in order to let the rest of the world come up to a sustainable level.

3. Not everyone aspires to what is sometimes called a 'Western lifestyle'. Millions of people around the world prefer different ways of living, and would probably choose to use rather less than 13,000 KWh per year. However, as I said in the Introduction, I'm not going to promote energy solutions that aren't based on fairness. I want us to envision a better world – a world where energy is not just the preserve of the richest. If we're going to harvest the natural energy of the sun, the tides and the Earth to improve our lives, then these gifts should be available to all the people of the planet. For this reason, it's important to show that it is possible to generate enough energy renewably for everyone to have access to fridges, buses, hospitals, DVD players, hot running water, organic denim jeans and takeaway curries *if they want them*. Otherwise, we'd be suggesting a global renewable energy system based on inequity and an unfair carving-up of the world's energy resources. This wouldn't just be morally wrong, it would also be incredibly difficult to achieve because some of the world's most important campaigners for clean energy are people who live outside the 'Western lifestyle' bubble (see Chapter 11). So it's important to ensure that the same level of energy is potentially available for everyone, even though lifestyles that use less energy than 13,000 KWh per year would still be very much available for those who prefer them!

OK. So we now have a rough figure for the amount of energy that we need for a decent life for everybody, and it looks low enough for us to meet with existing renewable technology, without stretching any of it too far. But there are still some really important questions to answer: can our renewable supply give us the right *kinds* of energy, in the right places and at the right times? What about the energy and materials required to manufacture all of the wind turbines, solar panels, rail tracks and electric buses required for this shift to a clean-energy future? Where will that come from, and what impacts will it have? We'll look at these questions in more detail in Chapter 10.

1 nin.tl/IEAtextbase **2** nin.tl/BPoutlook **3** nin.tl/Shellscenarios

10 A 100-per-cent renewable world?

> *'What happens when renewable energy runs out?'*
> – Victoria Ayling, candidate for the UK Independence Party in Great Grimsby, quoted in the *Grimsby Telegraph* in February 2015

The aim of this chapter is not to present some kind of 'perfect' model or blueprint of what a renewably powered world should look like. The aim is simply to answer the question: is it possible to give everyone a good quality of life using existing renewable technology? If it is indeed possible, then what are the barriers to making this happen, and what are the main risks, limits and crunch points that we need to watch out for along the way?

In Chapter 9, we saw that, when it comes to the raw numbers, it appears perfectly possible to generate enough renewable energy every year to provide nine (or even 10 or 11) billion people with a good quality of life.

However, these broad numbers leave us with a lot of unanswered questions. It's all very well to say that we can supply 13,000 KWh of clean energy per person per year – but can we get that energy in the right form (electricity, heat, liquid fuels), in the right place, and at the right time? How can we make sure that variable energy sources like the sun, the wind and the waves are providing us with power at the moments when we need it most? And what will be the environmental impact of building all of these new solar panels, wind turbines and tidal generators (not to mention electric vehicles, efficient homes and energy storage systems)?

The right energy

In 2013, humanity burned through 155,000 TWh of energy. Here's where it all came from:

Figure 10.1: Where the world's energy came from in 2013

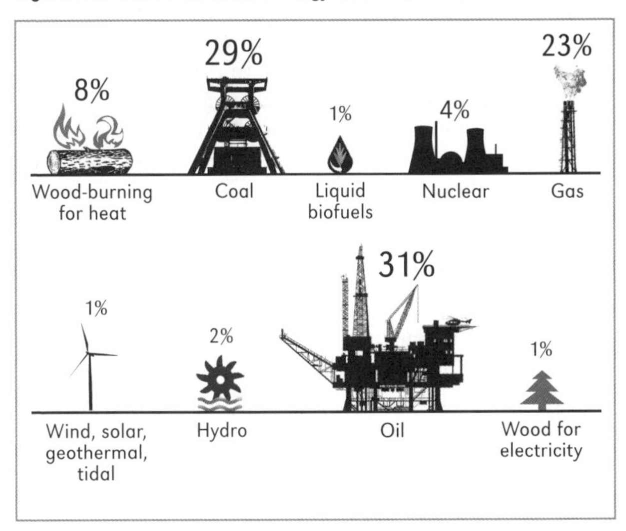

...and here's where it all went:

Figure 10.2: Breakdown of global energy use in 2013

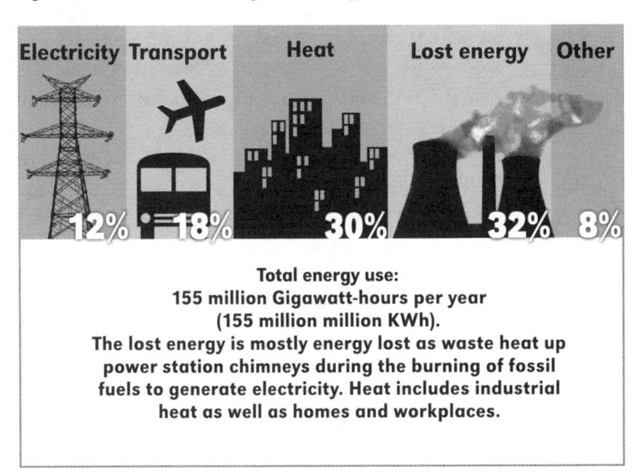

There are several important things to notice here. First, the biggest category is 'lost energy'. This is mostly energy lost as waste heat disappears up power-station chimneys during the burning of fossil fuels to generate electricity. Using coal and gas for power is a wasteful business.

Next, we need to note that when it comes to 'final energy consumption' – that is, the actual end-use of the energy after losses – the biggest segment is heat, making up 44 per cent (after losses). This includes industrial heat as well as warmth for homes and workplaces, and refers to heat obtained directly from burning coal, gas, oil and wood, or captured straight from the sun.

The second-biggest chunk of energy use is transport fuel, making up 27 per cent of global energy use (after losses). This is made up of around 95 per cent fossil oil and 5 per cent liquid biofuels.

Electricity makes up only 18 per cent of final energy use. By this point, you might have spotted a problem. A lot of our potential sources of renewable energy – solar PV, wind, wave, hydro, and tidal power – can only supply us with electricity. However, electricity currently supplies less than a fifth of our energy. Global society is set up to run mostly on direct heat and liquid fuels, with electricity as a much smaller (though important) segment.

As we've seen, we can get a good chunk of renewable heat from rooftop solar water heating, concentrating solar power, biogas from waste, and geothermal energy. We can also get a small amount of liquid fuel from waste cooking oil. However, these sources won't be enough by themselves to provide the quantity of heat and transport fuel we need. We're going to have to make some other changes too.

Electrification

In order to match up our energy needs to our energy supplies, the future will need to be much more electric.

As we saw in Chapter 4, ground-, air- and water-source heat pumps can be a great way of transforming a small amount of electricity into a larger amount of heat. They can play a major role in replacing the direct heat we currently get from fossil fuels with clean heating powered by renewable electricity.

Electric vehicles also have a large part to play here. Electrified railways are already a familiar technology, particularly in Europe and Asia; battery-powered cars and buses are also beginning to find a foothold.

Because much of the energy in petrol and diesel is lost when it is burned (as waste heat or as sound), electric vehicles also use significantly less energy than fossil-fuelled cars per kilometer travelled. Electric cars do require more energy to manufacture (because making batteries is pretty energy-intensive), but even when this is taken into account electric cars require 0.5 KWh per kilometer travelled, compared to 0.8 KWh per kilometer for fossil-fuelled cars.[1]

However, even though electric cars are more energy efficient, simply swapping all our existing internal combustion cars over to electric would still leave us with an energy demand too high to be met by renewables. We also need a serious shift away from mass car use towards reliable, affordable (and electric) public transport, as well as walking, cycling and car sharing. For example, around 80 per cent of the overland distance travelled by the average UK resident is currently done by car; our model of a more sustainable 13,000 KWh per year involves halving that to around 40 per cent,[2] with the rest being shifted onto buses, trains, bicycles and feet. It's worth remembering, of course, that 90 per cent of the world's population do not own a car, and so universal car sharing and public transport would represent for most people a great increase in mobility, rather than the 'loss' of car ownership.

These changes would have a host of other benefits as well – reducing road accidents, congestion and impacts on wildlife as well as increased mobility for poorer people and healthier populations. It would also be vital for minimizing the amount of new materials we'll need for our transition to a sustainable world; it obviously takes much less steel, copper and plastic to make one electric bus rather than 80 electric cars.

Tricky transport

While the great majority of transport could be electrified, there are a few types of vehicles that require too much energy, over too long a period, to be run on current battery technology. The most important of these are long-distance trucks, large container ships and aeroplanes.

The first thing to say about all three of these transport types is this: in a more sustainable future we'd need to use less of them per capita, compared with current usage in Europe, Australia and North America. A shift towards more local and regional production of food and goods, a general reduction in consumer junk, and a shifting of goods transport from road and air onto rail and water would reduce the energy needed for freight in the wealthier countries. This would allow the rest of the world to increase to a fair and sustainable level, keeping us within our 13,000 KWh/person/year target.

Also, of course, there is a real possibility that improvements in batteries and electric motors might eventually allow the electrification of some of these tricky bits of transport. However, for the moment we're trying to answer the question of whether *existing* renewable technology can give everyone a good quality of life, so we'll leave aside those future possibilities for now.

Many long-distance trucks are too heavy to be battery-powered. Some could potentially be run on renewable hydrogen (see box p136) or waste gas; the Zero Carbon Britain report suggests that synthetic liquid fuels could have a major role here (see box on page 139).

Container ships are often forgotten when we think of global energy use. Around 10 billion tonnes of cargo –

90 per cent of international trade – was moved between nations by sea in 2014, with several billion more being shipped within countries. International ships are powered almost exclusively by fossil fuels and produce around two per cent of global greenhouse-gas emissions, more than almost any country.

Could one answer be a return to the age of sail? Sail transport initiatives have been springing up around the world. Some have a local focus, like the Salish Sea Trading Co-operative, a sailor-owned project transporting local products around Puget Sound in the US state of Washington. Based in Seattle, they use exclusively sail-powered vessels but are fundraising for a purpose-built hybrid boat with a back-up electric engine. If replicated widely, this kind of initiative could cut a significant chunk out of coastal transport emissions.

Others have their sights set on international trade. The British-based B9 Shipping has designed a 4,500-tonne-capacity container ship, powered by 55-meter high sails with a back-up engine fuelled by waste gas, which could be a serious contender for cleaner global trade. Ultimately, the company hopes to build ships with a capacity of 25,000 tonnes.

Similar schemes are springing up around the world, under the umbrella of the International Windship Association.[3] These projects are using different mixtures of sail power, waste gas, renewable hydrogen and solar-charged electric engines.

Another potential solution for long-distance freight could be airships. Modern airships (filled with helium rather than the flammable gases of the past!) are much more efficient than planes, and can be run on renewable hydrogen fuel. They are also slower than planes, so more suited to freight than to passenger travel (passengers on a three-day trip over the Atlantic would need cabins, taking up extra space and making this form of travel more expensive).

Renewable hydrogen

Water molecules can be split with an electric current, converting water into hydrogen and oxygen gases. This process is called electrolysis.

Electrolysis can be a doubly useful process when it comes to renewable energy. It can transform electricity from solar, wind or tidal power into hydrogen, which is a very useful fuel for certain purposes (such as long-distance transport). It can also be a good way of capturing surplus energy at times of high supply. For example, wind turbines often generate a lot of power in the middle of the night, when demand is low. Using that electricity to generate hydrogen is one way of making sure that wind power doesn't go to waste.

However, switching energy from one form to another inevitably leads to energy loss. When electricity is used to make hydrogen, about 30 per cent of the energy in the electricity is lost along the way. For this reason, we should only use hydrogen for those specific purposes where we don't have a better alternative.

Industrial heat

This could be largely electrified, but there are some high-temperature processes where it can't quite do the trick yet. Countries with a decent amount of sun should be able to plug much of this gap with concentrated solar heat;[4] other options include renewable hydrogen, geothermal heat, and something called synthetic gas that I'll explain below.

In the right place, at the right time

The sun doesn't always shine when we want it to, and the wind doesn't blow for us on cue. Even if we site our solar panels, wave generators and wind turbines in the best possible spots, and generate enough power overall, there will still be times when supply is higher than demand, and vice versa.

We can reduce this problem if we time our energy demand to match the times of high supply – for example, by aiming to use electrical appliances at times when there's surplus power on the grid. We do a small version of this kind of 'demand management' at our house – when the sun is shining on our solar panels, we rush to load up the washing machine or do the hoovering. This could be made easier by smart metering technology, that could make these decisions automatically for householders (see box).

Another useful option is energy sharing or trading between countries or regions. This is already common in northern Europe, where – for example – Denmark sells surplus electricity to its neighbors when its many wind turbines are doing well, and buys spare hydropower from Norway when the wind is low.

As well as managing demand and sharing energy across borders, we can also store some energy, using batteries, compressed air storage or pumped hydro (Chapter 3). However, at the moment none of these technologies are quite good enough to solve the problem completely (although batteries are rapidly improving, and heat storage in CSP plants may also help – see Chapter 1). In a 100-per-cent renewable world we will still need some kind of back-up power that we can switch on and off whenever we like.

Electricity generators that can provide a large amount of power on demand at short notice are known as 'rampable' or 'dispatchable' energy sources. At the

Smart grids: automatic for the people?

In a renewable future, we'll need to get better at using the right energy at the right time. This is where 'smart metering' and 'dynamic demand control' come in. These technologies can detect when there's surplus power in the grid, and time the cooling cycles of our fridges and freezers, and the running times of our washing machines to match it. This idea raises understandable concerns about data security, but would create significant energy savings and may have other benefits. Industry analyst Jeremy Rifkin foresees a new collaborative energy commons, with homes, businesses, schools and hospitals connected by a 'smart grid' that shares energy automatically back and forth to where it's needed. If more buildings have their own renewable power generation (such as solar panels) and energy storage (such as batteries), then this interconnected web of small producers and consumers could take a significant load off our national electricity grids, reducing reliance on centralized power stations.[5]

moment, natural gas and large dams are typically used for this purpose. In a 100-per-cent renewable world where storage isn't yet good enough to plug the gap (and where we want to reduce our reliance on megadams), there are two broad options available:

- In many countries, geothermal electricity could provide a realistic source of rampable back-up power.
- Countries without much geothermal potential could instead try combining renewable hydrogen with woody/grassy fuels to make a high-energy biogas to burn for back-up power (see box). However, such fuels would need to be organically grown under local control, detached from deforestation or land grabs, and very carefully managed.

So, in theory, it's already possible to use 100-per-cent renewable (and dam-free) energy for back-up generation. In practice, however, there are still some serious obstacles to face. The most important ones are to do with the social and environmental impact of using woody and grassy fuels, and the challenges of ownership of, and access to, productive land around the world. We would need every country where such crops are grown to have strong legislation in place to ensure that such crops are produced locally, organically and without the exploitation of people or the environment.

We'd also need small-scale farmers the world over to have greater sovereignty over their lands and livelihoods, so we could be confident that any energy crops grown around the world were being produced by choice, with farmers able to prioritize the production of food for their communities and only growing fuel crops as a voluntary add-on, for extra income. We'd need to know that no-one was being forced into producing energy crops against their will, and that no land was being grabbed by energy companies or governments for the purpose. We'd also need to be sure that no forests or wetlands were being cleared to produce fuel crops; instead, we'd need to use only sustainable wood

Sustainable bioenergy?

In its 2013 report *Zero Carbon Britain 2030*, the Centre for Alternative Technology (CAT) in Wales paints a picture of how the whole of the UK could be powered by renewable energy in 2030. In this model, it suggests that one way of balancing the UK's energy needs could be to use the surplus electricity generated on windy days to make hydrogen; this could then be combined with woody and grassy crops to make two very useful products: liquid fuels (via a chemical reaction called the Fischer-Tropsch process), and methane gas (via a similar reaction called the Sabatier process).

These processes require significantly less plant material than making cellulosic ethanol or biogas directly from woody and grassy crops. The hydrogen is providing a lot of the energy for the fuel, which means that much less crop material is needed. This also helps to prevent excess electricity from particularly windy (or sunny) periods from being wasted.

However, we'd still need *some* crops to do this. *Zero Carbon Britain* suggests that they could be locally, organically grown and carefully managed; it proposes that the necessary land could be freed up in the UK by a shift from livestock farming to vegetables and grains (which can feed more people using fewer hectares).[6] In this scenario, relatively poor soils currently used for grazing sheep and cows could be turned over to producing hardy miscanthus grass or coppiced willow, without creating an 'ILUC' effect (see Chapter 6).

On a global scale, meanwhile, if the many households currently burning wood fuel had access to cleaner, more efficient methods of heating and cooking instead, this would free up a significant amount of wood which could – if harvested fairly and sustainably – provide a source of feedstock for renewable liquid and gaseous fuels.

supplies made available by a reduction in wood-burning elsewhere, or ex-livestock land freed up by a shift away from meat and dairy in the world's diet.

The fact that these political, social and economic changes might also allow us to produce woody and grassy fuel crops in a sustainable fashion simply gives us an extra reason to demand the fairer, more sustainable food system that we should be calling for already.*

* See *The No-Nonsense Guide to Climate Change* (2010) for more details.

Meanwhile, the back-up system we're talking about would only need to be used at moments of particularly high demand (for example, when a large percentage of the country switches the kettle on at half-time during a major televised sporting event, or turns on their air-source heat pumps at the same time on a chilly morning), or at times of extended low supply (for example, a few days of unusually low wind combined with cloudy skies across a whole country or region). The Centre for Alternative Technology estimates that only 3 per cent of electricity use in a fully renewable UK would need to come from back-up generators; the remaining 97 per cent would come from wind, tidal, wave and solar generation, either directly or via energy storage.[7] The Zero Carbon Australia project found that – with the help of heat storage in CSP plants – only 2 per cent of electricity demand would need to come from rampable backup power in a completely renewable Australian grid.[8]

Electricity grids that include large amounts of variable renewable supply are currently working just fine in countries like Spain, Denmark and Germany. National-grid administrators are already adept at keeping the lights on despite large variations in national energy demand, if they have the necessary information and enough rampable sources to choose from. These days, weather forecasts are surprisingly accurate a few days in advance, allowing grid co-ordinators to plan around changes in wind and solar supply.

This means that the variability of renewables does *not* need to be a barrier to our cleaner energy future. We can get on with the transition to an almost-entirely renewable grid, and work out the best methods for back-up power as we go along. One option for countries without access to geothermal backup energy would be to keep some natural gas (and/or big hydro) stations, usually switched off but on standby, ready to provide that 2-3 per cent of backup power for the time being. These could then be shut down permanently once we've sorted out better energy storage, and/or a fair and sustainable method of supplying energy crops.

Finding a flight path

OK, it's time to talk about flying. First things first: we have to let go of the idea that anyone can just jump in a plane and fly anywhere they like, whenever they like. Not only would it be impossible to provide enough renewable energy for everyone in the world to do this, it's also never really been true. Over 90 per cent of the world's population have never been on a plane. Even in an aviation-happy country like the UK, half of the population do not fly in any given year.[9] A quarter of US citizens have never flown anywhere.[10]

Even amongst those who do fly, the majority of flights are taken by the wealthiest minority. Some 79 per cent of holiday flights from the UK are taken by people earning more than the median income. The poorest third of the population take just 11 per cent of flights.[11]

So a decrease in flights in wealthy countries wouldn't have much effect on the majority of people, who rarely (or never) fly in any case. However, we still need to know: is it possible to do *any* flying sustainably? It takes a large amount of concentrated energy to hurl a sizeable metal vehicle into the sky, keep it up there for hours and bring it safely back down again. A jumbo jet flying from New York to London uses around 840,000 KWh of energy each way – enough to provide 65 people's energy needs for a year. Trying to store that much power using existing battery technology would make the plane too heavy to leave the ground.

Hydrogen-powered planes are also a no-no. Burning hydrogen produces water vapor; this isn't a problem in ships, trucks and low-flying airships, but pumping water into the air at the high altitude used by planes would be disastrous for the climate, because at those heights water acts as a powerful greenhouse gas.

Aviation-industry spokespeople tell us that fully

electric flights might be a possibility by 2050, while also stating that flights would increase sevenfold in the same period (with most of that growth happening within the richest minority of the global population). Expanding fossil-fuelled aviation for the next 35 years would be disastrous for the climate,[12] and a transition to electric planes some time after 2050 would be far too late to help.

Perhaps, with better incentives and resources, 100-per-cent electric passenger flight might become a reality sooner than 2050; even so, we should leave that aside for the moment because (at the risk of sounding like a stuck record) we're trying to find out if a fully renewable world is possible using *existing* technology. Once again, the Zero Carbon Britain report has a suggestion. It recommends using surplus renewable electricity to make hydrogen and then adding this to woody and grassy crops (see box p.139) to make high-energy liquid fuels for aviation. This would allow everyone in Britain to take the equivalent of one long-haul flight every few years – about the current amount for someone on an average income in the UK.

What if the whole world had access to this much aviation? Let's assume that a third of these flights could be shifted onto airships fuelled by renewable hydrogen, with the rest being powered by synthetic liquid fuels. Let's also assume that another large chunk of synthetic liquid fuel would be required for long-distance road freight and shipping, based on the numbers in the Zero Carbon Britain report. All of this energy is included in our 13,000 KWh/person total (about 350 KWh/person for flights and 750 KWh/person for road freight and shipping).

In total, this would require around 330 million hectares of land. If we also used fuel crops (combined with renewable hydrogen) for our back-up electricity supply and some industrial and commercial heating, this would require a further 240 million hectares – a total of 570 million. This is, in fact, less than the estimated 800 million hectares currently used to provide wood fuel, so, if organized in a fair and sustainable way, this transition could result in a reduction in the amount of land required for growing energy worldwide.

However, once again it is vital to state that this path could be disastrous without significant changes to our political and economic systems. We cannot have a future where large corporations and governments seize land from local farmers or cut down rainforests in order to grow fuel crops. Instead, we want a system where farmers can choose to grow energy crops as an extra source of income if they wish, boosting rural economies without taking control out of the growers' hands.

In these circumstances, we would expect many people to choose to grow food to feed themselves and their communities rather than fuel crops. It would only need a minority of farmers in each country to decide to grow fuels (probably alongside other crops) in order to reach the quantities needed for that much flying, and many will already be growing wood for fuel in any case. Others will already be growing grass for animals, which could be switched to a grassy fuel crop if meat consumption falls.

We can also hope that improvement in renewable technologies (such as better vehicle batteries and energy storage methods) will mean that we'll need far less plant-based fuel than in this model.

Another possibility would be, of course, to fly less. Will people even want to fly that much? Even one long-haul flight every few years is a huge increase for 95 per cent of the global population. The point is that it's possible – but it may not be desirable. Many people across the world may well decide that they'd prefer to have fewer flights and thus avoid the need to grow fuel crops. This again shows the importance of securing local people's sovereignty over agricultural production before going down this route; without it, we'd be sure to see people across the world being forced to grow fuel to feed the planes of the rich.

An even bigger user of liquid fuels in our 13,000-KWh lifestyle is long-distance road freight and shipping. However, this could of course be reduced if we, the peoples of the world, decide that it makes more sense to produce things even more locally and regionally than in the Zero Carbon Britain model, and thus need less land for fuel crops. Countries that are less addicted than the UK to long-distance freight will probably find this easier to achieve.

Down to earth

To sum up: renewably powered flying is indeed possible with existing technology, but with a cargo bay's worth of caveats.

However, flights currently use just two per cent of global energy. This doesn't mean it isn't a problem; we definitely need to minimize fossil-fuelled aviation so it doesn't undo all our good work reducing fossil-fuel use elsewhere (especially from a justice perspective, as aviation benefits only a tiny minority of the world's population, and also because jet fuel burned at high altitude has an increased global-warming impact). However, there's a strong argument to say we should get on with the rest of the clean-energy transition, and just keep flying to a minimum until we've sorted out the alternatives. Maybe we'll all get used to slower travel, by train, boat and airship, with heavier-than-air flights reserved for emergency use. Maybe a future less driven by consumerism will allow us all to work a bit less and have more time for slower travel. Maybe videoconferencing technology will become good, cheap and widespread enough for us to feel like we're visiting people around the world from the comfort of our homes and offices. Maybe there'll be a technological breakthrough in electric planes or synthetic fuels, or global land reform that makes fuel crops a viable option. In any case, I hate to say it, but figuring out ways for lots of people to travel fast isn't a priority compared to providing everyone with comfortable homes, good food, healthcare, education and secure livelihoods. This is where we need to be focusing our renewable-energy efforts.

Building the future: how much material will we need?

Wind turbines and solar panels aren't made from pixie dust. Once they're up and running, renewable generators require little in the way of inputs, but building them in the first place takes a whole heap of materials.

I've taken recent research into the amounts of materials needed for different types of renewable energy,[13] and tried to work out: how much material would be needed for a transition to a 100-per-cent renewable world, where everyone had access to 13,000 KWh of energy per year? You can see my full calculations online [nin.tl/materialrequirements], but the headline result was that we'd need something in the region of five billion tonnes of iron, cement, aluminum, and copper to build that much energy generation equipment (solar panels, wind turbines, geothermal plants, heat pumps, solar water heaters and so on).

On top of that, we'd need another two billion tonnes or so to build a whole new fleet of electric vehicles (prioritizing buses, trains and bicycles). A worst-case back-up scenario – where we built enough reserve power stations to cover all of our solar and wind generators in case they all stopped generating at once – would need another half a billion tonnes.

However, in order to make this transition sustainably we'll also need to dismantle all the existing dirty infrastructure, from oil pipelines to petrol cars. Recycling this will give us back at least a billion tonnes of iron, aluminum and copper to feed into our transition, reducing the total demand to 6.5 billion tonnes.

Spreading this 6.5 billion tonnes over 25 years gives us a material demand of 0.25 billion tonnes per year. The world currently consumes around five billion tonnes of these materials per year, so we'd need to redirect around five per cent of current global production of iron, cement, aluminum and copper into our sustainable transition.* Within that total, copper will be a particular stretch, requiring around a third of current annual production to be switched over to building clean-energy technology.

This is a serious undertaking, but does appear to be possible, especially when we consider the material use that our transition will avoid. For example, in a 'business as usual' future we'd be using two billion tonnes of materials to double the world's current car fleet. In our cleaner-energy future, we can divert that material into sustainable technology instead.

Once we've built everything the first time the metals can be recycled indefinitely, meaning that we're talking about a short-term burst of new material use to get everything installed, from which point onwards we'll be able to get most of what we need from recycling the old turbines, panels and so on. But that short-term burst will still require a lot of mined materials. Even if we're diverting them from other uses, they'll still need to be dug out of the ground. What will be the impact of this?

There is no such thing as zero-impact mining; it is one of the most notoriously destructive, poisonous and corrupt industries in the world. The final amount of raw material produced is just the tip of the extraction iceberg; every tonne of metal or cement requires many more tonnes of rock and ore to be hauled out of the ground in the mining and production process. Our 6.5 billion tonnes of copper, aluminum, iron and cement will require 60-80 billion tonnes of real-life extraction.

However, we need to look at the other side of the equation too. Phasing out fossil fuels over the next 25 years will mean a huge reduction in the amount of oil, coal and gas extracted over that period. Based on IEA

* At current extraction rates, there are more than enough of all these materials in proven reserves to last for decades to come; once extracted, the metals can in theory be recycled indefinitely.

projections, shifting to 100-per-cent renewables would avoid the need for around 230 billion tonnes of fossil fuels between now and 2040. Coal, tar sands and heavy oil, like metals, require the extraction of large amounts of extra rock and earth; when all this is added in, our transition would prevent 1,850 billion tonnes of fossil-related extraction up to 2040.

So even if we needed the full 80 billion tonnes of extraction to build our renewable future, we'd still be creating a huge reduction in the amount of destructive extractive industry taking place worldwide.

As well as looking at the big mining totals, we need to think about the availability of certain hard-to-find materials. Again, I looked at this in some detail in an online article [nin.tl/materialrequirements] and found that most of the rarer minerals (like dysprosium, neodymium and cobalt) needed for things like solar panels, wind turbines and electric cars could be replaced with more common alternatives, and so shouldn't be a barrier to our transition.

However, there is one element that could pose a problem: lithium, which is used to make the rechargeable batteries in electric vehicles. A mass rollout of electric cars could exhaust proven lithium reserves within 100 years – not counting the extra lithium that might be needed for improved electricity storage systems in homes and communities. This means that humanity should be able to obtain enough lithium to make the initial transition to an electrified transport system, but to maintain it beyond the second half of the century we'll need to either get very good at recycling it, find more supplies, or find safe and affordable ways to extract lithium from the oceans (where it is abundant, but dispersed).

There's another serious issue here. This is one of those moments where it's easy to slip accidentally into a colonialist mindset, when referring casually to 'reserves' of minerals 'available' to the world. Whether or not those materials are dug out of the ground should not be a decision for someone like me, a white guy typing on a computer in Europe; it should be up to the communities that live in the area concerned and would be affected by the extraction. Although the quantities of lithium required for everyone in the world to have decent access to electrified transport are relatively small when compared to high-volume mined materials like iron or coal, the necessary mines would no doubt loom large in their local landscape. Most of the world's known lithium reserves are located in Bolivia and Chile. These are real places, inhabited by real people – including Indigenous peoples whose lives, livelihoods and culture are intimately bound up with the land they live on. Will it be possible to obtain enough lithium for an electrified world without trampling over the rights of local communities? If not, then we'll need to find a different path to our renewably powered future.

A qualified conclusion

The big question I've been trying to answer in the last two chapters is: 'can everyone on the planet have access to enough energy for a good quality of life, using 100-per-cent renewable energy?' The answer, as I hope has come across, is a qualified 'yes'.

It matters that the answer is 'yes'. We need to be able to talk positively about the future. We need to be able to tell people that the world can leave fossil fuels behind without also abandoning essentials like comfortable homes, healthy food, decent healthcare, public services and education, along with enough transport to stay well connected with society and 'luxuries' like arts, culture, and entertainment. It's vital that when we talk about this future, we do so in the knowledge that it can be equally accessible to everyone; our vision doesn't rely on some staying energy-poor so the rest can be energy-rich. It also matters that we can do this – just about – using existing technology, without needing some sort of uncertain future breakthrough to make it a reality.

It's good to know that we can say these things, and I encourage you to do so, loudly and often. We won't be able to leave fossil fuels behind unless enough people demand it; and people are less likely to demand change if they don't believe there's a better alternative. However, it's also important to remember that this is a qualified 'yes'. It comes with a big fat warning, and then a number of ifs and buts.

A big fat warning

This book does not contain 'the' correct blueprint for a renewable world. All I've done here is make a rough sketch of how, in theory, one example of a low-energy, renewably powered version of a 'modern' quality of life could be technically possible for everyone on the planet. This is just an example, to make a point. Ultimately, it isn't up to me, or you, or any one person to decide what sort of life different people will choose to lead in a post-carbon world. We do need to stop burning fossil fuels, and there are limits to the amount of renewable energy we can sustainably generate. However, what we, as people and communities, choose to do with that new, clean energy is limited only by our imagination. Moving beyond fossil fuels will need more than the single vision put forward in this book; we need a million overlapping visions all over the world, to match the needs, dreams and desires of different people in different places. A million different visions, united by a common goal of a fairer, safer planet.

Ifs and buts

We can power the world renewably, if...

- The wealthy minority reduce their overconsumption so that the rest of the world can come up to a sustainable level. This includes the purchase of consumer junk as well as direct energy use.
- Travel is mainly carried out by public transport, walking and cycling, with cars used only when

absolutely necessary, mainly through car-sharing schemes. This might be annoying for the 10 per cent of the global population who currently own cars, but for everyone else it'll be a big increase in mobility.
- Nearly all transport is electrified.
- All buildings are properly 'weatherized', to stay warm in cold seasons and cool in hot seasons with minimum energy use.
- We support small farmers and Indigenous peoples around the world to take control over their lands and forests. These changes would preserve forests and livelihoods, feed more people, and allow more communities to grow and distribute sustainable energy crops without fear of exploitation.
- The world (and especially the rich world)'s diet shifts away from meat and dairy, back towards vegetables and cereals, freeing up land both to feed more people and to grow local energy crops.
- We keep some gas power stations as an occasional (2-3 per cent) back-up to deal with the variability of wind and solar power, until we have enough rampable energy from geothermal, electricity storage and/or sustainable local energy crops to fill the gap.
- We minimize heavier-than-air flights and find alternative, slower ways to travel until sufficient sustainable energy crops (combined with renewable hydrogen to make synthetic liquid fuel) are locally available, or until fully electrified flight becomes a reality.

We can power the world renewably, but...
- There are limits to how much we can generate. The model described here falls within those limits (assuming that the maximum generation estimates from various renewable-energy experts cited in this book aren't wildly wrong). We are pushing at the edge of some of these limits, though, particularly with regard to annual copper use and total lithium requirements.
- No form of energy is impact-free. Renewables may

be less destructive than fossil fuels but, if we don't roll them out carefully and sensitively, they could still create serious problems, from toxic mining for the raw materials to wildlife impacts at the generation sites.

- Some of the potential generation numbers in Figure 8.7 make it look as though we're only using a small fraction of the available energy in our example, especially from sources like solar and geothermal power. There are very good reasons for this. Beyond the first 15,000 TWh/year of rooftop panels we would need to use solar farms, which take up land space that is valuable for other purposes, from farming to wildlife habitat. Too much reliance on CSP from deserts would require the export of energy from centralized generators in specific countries, increasing the risk of corruption and colonial-style exploitation. Advanced geothermal technology is still in its early stages, so we don't yet know if it will cause significant local pollution or seismic impacts. Plus, of course, we need to keep our raw-material requirements to a minimum, so we don't run out of important components or stimulate disastrous new mining projects. For all these reasons, we can't just look at the numbers and assume 'ooh, we've got loads of potential renewable energy, let's build build build!'

Bringing down the barriers

It should be clear from the list above that a shift to a 100-per-cent renewable world is about far more than just swapping technologies. We need significant social, political and economic change in order to make this work.

This isn't a small thing to ask! It's something I covered in a lot more detail in the *No-Nonsense Guide to Climate Change*, where I concluded that these social and political shifts are *theoretically* possible in the time available, but will require a serious increase in public engagement and action on these issues. There are signs that this is beginning to happen, but not yet on the scale required.

We'll look at this again in Chapter 12.

The good news is that most of the changes we need should be positive overall. We'd be creating a world where everyone has access to the energy they need; a world of comfortable homes and decent public transport, a world where billions escape from fuel poverty while the richer minority break out of the consumerist rat race and reconnect with the things that genuinely make life worth living.

However, there are also major barriers standing in our way. One of the biggest is the issue of who currently owns, controls and profits from our energy supply. In Chapter 11 we'll ask the question: who has power over power?

1 nin.tl/batteryvehicles **2** twoenergyfutures.org **3** wind-ship.org **4** nin.tl/renewgamechanger **5** nin.tl/electricalefficiency **6** zerocarbonbritain.com **7** Research by Dan Quiggan of Demand Energy Equality suggests that the Zero Carbon Britain model underestimates the amount of back-up power needed to deal with peaks in electrified heating. This means that temperate (and colder) countries like Britain might need a slightly different set-up: a mix of heat pumps and district heating from Combined Heat and Power (CHP) plants fuelled by sustainable local biomass or synthetic biogas, as described by the 'Thousand Flowers' model developed by researchers for the (also UK-based) Realising Transition Pathways project. CHP plants make electricity but also capture the waste heat from burning the fuel, and use it for local heating. This would require roughly the same amount of back-up fuel as in ZCB, but this biogas/biomass would be used in a more distributed fashion: realisingtransitionpathways.org.uk/realisingtransitionpathways/publications/Working_papers/RTP_WP_2013_5_Barton_et_al_-_Comparing_scenarios_and_technology_implications.pdf **8** skepticalscience.com/Zero-Carbon-Australia-2020.html **9** UK Civil Aviation Authority **10** nin.tl/USairtravel **11** UK Civil Aviation Authority and Office of National Statistics **12** nin.tl/1FvRttA **13** Edgar G Hertwich et al, 'Integrated life-cycle assessment of electricity-supply scenarios confirms global environmental benefit of low-carbon technologies', PNAS, 2014, nin.tl/electricproof; MA Delucchi et al, 'An assessment of electric vehicles', Royal Society, 2013, nin.tl/vehicleselectric

11 Whose renewable future?

> *'I guess we'll just move into insulation and renewable energy.'*
> – A representative of the large utility company British Gas, when asked by a campaigner what the corporation would do if the anti-fossil-fuel movement was successful

In January 2015, the energy researcher Jeremy Leggett made a bold claim. He told the *Guardian* newspaper that we should expect a major oil firm to turn its back on fossil fuels soon and shift to renewable energy. 'One of the oil companies will break ranks,' he said, 'and this time it is going to stick.'[1]

Leggett points to the collapsed oil price, the falling costs of renewable-energy generation and potential government action on climate change as key factors that could persuade an oil corporation to jump ship. His comments were excitedly shared online by anti-fossil-fuel campaigners.

But hang on a minute. Would this really be good news? To avoid catastrophic global climate change, we need to leave at least 80 per cent of known fossil fuels in the ground, and renewable energy will have a major part to play in that. But do we want our new, clean energy system to be owned and operated by the same corporations that have got us into our current mess? Do we trust the likes of BP, Exxon and Total to develop renewables in a fair and sustainable way?

To answer this question, we don't need to look far. All over the world, companies and governments that have grown rich on our current fossil-fuelled system are doing their best to slow down, interfere with, co-opt and control the growth of renewable energy. We need to fight back with a different vision: of a democratically controlled, people-focused clean-energy system built from the grassroots up.

A renewables revolution?

The year 2014 felt like a big step forward for renewables. The amount of wind and solar power installed around the world grew by 15 and 32 per cent respectively. Solar electricity is now cheaper than the grid average in Spain, Italy, Australia, Chile, Germany, Brazil and at least 10 US states. UBS, the world's biggest private bank, told its investors that large, centralized power stations are on the way out in Europe, to be rendered redundant by rooftop solar panels and home energy storage in the next 20 years. Meanwhile, the governments of India and China have announced solar- and wind-power schemes large enough to send panic through the Australian coal industry, whose expansion plans were reliant on exports to those countries.

These could be early steps towards a better energy future. As we've seen in the last couple of chapters, it is technically possible for everyone on the planet to have enough energy for a good quality of life, using only renewable technology that already exists. However, this will only be possible if the wealthy minority – mostly in Northern countries – stops overconsuming energy, so that everyone else can come up to a fair and sustainable level. This isn't currently happening; instead, barring a blip for the 2008 financial crisis, total energy use in OECD countries has been steadily rising.[2]

Avoiding runaway climate change will also require our new cleaner energy sources actively to replace fossil-fuel generation, not just add to it. There's little point installing a solar-powered radio in a diesel-fuelled SUV. The current blossoming of renewable energy has dented coal use in a few countries – notably the US and China – but has so far failed to make much impact on a global scale. Between 2010 and 2013, the annual production of renewable energy grew by around 0.5 million GWh to 20 million GWh per year; in the same period, annual fossil-fuel use grew by 8 million GWh – 16 times faster – to reach 128 million GWh/year.[3]

India: a beacon in the gloom

In the Bihar region, more than 80 per cent of the population have no access to electricity. Dharnai village is no longer part of that statistic. In 2014, after 30 years of reliance on wood, cow dung and kerosene, the lights came on in Dharnai thanks to a solar-powered microgrid installed by Greenpeace India. The village's 450 households, 50 businesses, 60 streetlights, two schools and health center are now all connected to the grid, providing over 2,000 people with 24-hour access to electricity. The 100-KW system with battery storage is managed by a 20-person Village Electrification Committee, which has a 25-per cent-minimum quota of female members and representation from people of various castes. The Committee sets the tariffs, aiming to make them low enough to be affordable while still covering the operating and maintenance costs of the system, including the training and employment of villagers as engineers. nin.tl/indiamicrogrid

The trouble is, these necessary steps to a safer future – ramping down fossil-fuel use, cutting Northern overconsumption and sharing clean energy fairly across the globe – fly directly in the face of our current growth-based economic system. As writer and activist Naomi Klein puts it: 'What the climate needs now is a contraction in humanity's use of resources; what our economic model demands is unfettered expansion.'[4]

Renewable energy does offer us a (solar) ray of hope amidst the climate doom and gloom. As shown by the positive examples in the boxes in this chapter, these technologies give us the potential to build a new, decentralized, democratic energy system that meets the needs of the many rather than providing profits to the few. But there are powerful economic forces and vested interests lined up against us, ready to steer renewable energy in a very different direction.

What are we up against?

Let's not kid ourselves. The fossil-fuel industry's main response to clean energy is to try to squash it. Selling the highly concentrated energy in oil, coal and gas is far more profitable in the short term than the slow-release, distributed energy from wind or solar power – especially when you factor in generous government fossil-fuel subsidies, an international energy infrastructure already set up to use these fuels, and free rein to pour carbon pollution into the air at little or no cost. Whether it's funding pro-fossil politicians, forging cosy links with officials or pouring money into anti-renewable front groups, the big oil, gas and coal companies are working hard to keep society hooked on their highly profitable products, and prevent alternatives from getting off the ground.[5]

There are exceptions to this rule. If those alternatives can provide decent short-term returns or access to new subsidies without disrupting the existing energy markets, then the big players might be tempted to step in. This is why the likes of BP, Shell and Exxon have moved into liquid biofuels, and why major power plants like Drax in Britain are starting to mix large quantities of wood fuel in with their coal supply (see Chapter 6).

Industrial biofuels and wood-fired power stations – along with the continued destruction caused by large hydropower dams – provide perfect examples of what

Renewable generators: big and central vs small and distributed

This varies greatly depending on the type of technology.
For example (2013 figures):

Rooftop solar PV:	118 GW
PV Solar farms:	21 GW
CSP farms:	3 GW

So 83% of solar electricity is distributed on roofs, not in big solar farms

Compare this with hydro:

3,520 TWh generated from big dams vs 270 TWh (7%) generated from small hydro schemes

can happen if supposedly 'renewable' energy sources are exploited for maximum profit, without proper consideration for people and the environment. Energy crops and hydroelectricity may both be sustainable on a small, local, carefully managed scale – but the current profit-driven rush to turn food crops and forests into fuel is leading to hunger, land grabs and deforestation, while megadams threaten huge areas of natural habitat along with the homes, lands and livelihoods of hundreds of thousands of people (see Chapter 3).

These projects should act as a stark warning. Wind and solar power are still relatively small industries on a global scale, but are growing fast. These technologies are far less destructive than fossil fuels, but – as we've seen – that doesn't mean they're impact-free, especially if they develop to the scale we need for a fossil-free future. Will they be carefully manufactured in renewably powered workshops with strict respect for workers' rights and environmental standards, using largely recycled materials, and built as part of community-run, co-operatively owned and democratic energy schemes which benefit the communities where they are sited? Or will they be churned out in nightmarish sweatshop conditions, using minerals from exploitative mining projects and sited in giant energy parks on cleared rainforest land from which the residents have been forcibly evicted?

It could go either way. Renewables could transform our energy system, with solar panels particularly well suited for decentralized use: 85 per cent of today's solar panels are spread over millions of rooftops, with only 15 per cent in solar parks. Increased access to and control over energy could empower millions of people, improving lives and livelihoods and boosting the political and social influence of marginalized communities.

Unfortunately, the risks are also clear. Wind and solar generators require a significant amount of building

Scotland: hatching alternatives on the Isle of Eigg

The Isle of Eigg may be small (five by nine kilometers, population just under 100), but it has some big lessons for us all. Located off the west coast of Scotland, the island was owned by a series of absentee landlords through the 1970s and 1980s, whose neglect created serious problems for the island's residents. In 1997, the community succeeded in a bold plan to buy back the island for themselves. After achieving community control, the residents set up the world's first-ever 100-per-cent renewable electricity grid. Every home on the island is now powered by a mixture of wind, small hydro and solar power, with a battery bank that can store 24 hours of back-up, all managed and maintained by a community-owned company.

material and land space. We saw in Chapter 10 that a transition to sustainable technology could mean a significant spike in demand for raw materials that could have serious local impacts around the world if not carefully managed. Wind power, unlike solar, is far more efficient when built on a large scale; big wind farms typically require levels of capital investment that are out of the reach of community groups. They're more likely to be installed by governments or large utility companies such as E.ON. Some 75 per cent of all wind turbines are manufactured by just 10 companies.

The large solar plants proposed for North African deserts have been criticized by local activists, who have pointed out that schemes like the Tunur project in Tunisia are set up to export electricity to Europe, rather than serve the needs of local people. These projects are seen as being imposed by transnational companies, using local land, water and labor to grow their own profits and provide benefits to overseas consumers, while also helping to prop up oppressive regimes in countries like Algeria.[6] If carried out in the wrong way, large renewable projects could develop along the same neocolonial and racist lines as our current fossil-fuel industry, where the rights of Indigenous peoples around the world are trampled in the pursuit of 'cheap' energy for the industrialized nations.

Germany: taking back power

In 2013, there were two landmark referendums in the German cities of Hamburg and Berlin. Citizens in both towns, sick of being exploited and polluted by commercial power providers, put forward proposals for their city councils to buy back their respective local energy grids. This 'municipalization' of energy would bring local energy provision back under democratic control, and commit to a transition to 100-per-cent clean power.

Berlin residents voted overwhelmingly in favor, with 83 per cent backing the proposal. However, the turnout was four per cent lower than the total required for the vote to be legally binding (many local campaigners blame the city government, for not allowing the vote to take place on the same day as the general election, when turnout would have been higher). The story is not over yet, though, because a clean-energy co-operative called 'Berlin Citizens' Energy' is now vying to take over the grid, with some support from city authorities. In Hamburg, meanwhile, the municipalization vote was a success, and the transition to greater renewable supply is under way. Hamburg joins 170 German municipalities that have reportedly brought their electricity grids back under local control since 2007.[7]

Who has power over power?

As wind and solar technology gets cheaper – and if low oil prices and increasing climate regulation make fossil fuels less profitable – we can expect more and bigger corporate players to move into the sector, including existing oil and gas corporations.

Energy supply in many countries is already in the hands of privatized utility companies, thanks to decades of privatization driven by neoliberal Northern governments and institutions like the World Bank. This has led to rising energy bills and the continuing failure to supply grid electricity to harder-to-reach (and thus less profitable) rural communities. Around 1.3 billion people worldwide still have no access to electricity, while many others struggle to afford it.

There are vital battles still to be fought over the ownership, control of and access to renewable energy. Who will pay for and own the building materials, the factories, the technical knowledge, the site of installation, the equipment and the energy it produces? The more democratic control that can be exerted over each stage of this process, the greater our chances of creating low-impact, climate-friendly energy systems that supply affordable energy to all.

We also need the transition to be a fair one that retrains and transfers workers from the fossil energy sector; this will only happen if the voices of workers carry more weight in the process than the desires of the energy companies.

It's hard to imagine the big privatized companies voluntarily working to reduce energy consumption in the North; it's equally hard to envision them supporting a phase-out of fossil fuels as renewables expand, or policies to provide affordable energy to those most in need. These companies have been driving our civilization towards a cliff edge, and now they are eyeing up the keys to our shiny, new, renewably powered electric bus.

In an energy system built from numerous small producers rather than just a few big power stations, ownership of the distribution network also becomes more important. Smart networks could make it easier

Indonesia: coming on stream

Around 70 million people in Indonesia have no reliable electricity supply. While the national electricity supplier PLN struggles to extend the grid beyond the towns and cities, mountain villagers are beginning to take matters into their own hands. With the support of NGOs like the People Centred Business and Economic Institute (IBEKA), small-scale hydropower co-operatives are springing up around the country. Typically, NGO engineers assess the site and build the system in collaboration with the villagers; at the end of the process, the community takes ownership of the hydro turbine and the power it generates. The electricity is then distributed locally, with tariffs set by the co-operative to cover the maintenance costs of the system; villages with a grid connection can also sell surplus power to PLN, following a successful national campaign to make such purchases a legal requirement for the company.

for producers and consumers to share and use energy more efficiently, reducing the need for centralized power production; however, whoever owns and controls these networks will ultimately determine who can use them and at what price, and will also have access to large amounts of data about households' energy usage. The energy giant E.ON has recently split its business in two, with one half taking the fossil power stations and the other focusing on renewables and 'smarter' distribution networks; Google has also invested billions in smart home metering.

Will companies like these be content with government or household contracts to build these systems, or will they seek ways to extract ongoing profits from charging for network use or selling customer data? Like the internet, an energy-sharing network could be a space for collaboration and co-operation, or for corporate rent-seeking and control; the way it is set up, and who owns the infrastructure, will be key. The governments of Spain and the US state of Arizona, under pressure from energy companies' lobbying, have announced taxes on solar-panel owners as 'payment' for their connection to the grid; this may be a taste of things to come.

We mustn't forget the technologies that use energy, either. A transition to electric vehicles will be vital for ending our dependence on oil, but this needs to be accompanied by a serious shift from private car use towards public transport and cycling, otherwise electricity demand will rocket beyond a level that can be sustainably met by renewables (see Chapters 9 and 10). Most of the expansion of the electric vehicle market has so far been driven by car manufacturers (particularly Renault, Nissan, Tesla and Mitsubishi), supported by government incentives; this could explain why the number of fully electric cars on the road is expected to hit a million in 2015, while electric bus numbers lag behind in the tens of thousands.

The rollout of electric vehicle charging stations in

the US (and increasingly Britain) is being spearheaded by Tesla, the company run by 'playboy billionaire' Elon Musk (the inspiration for Tony Stark in the *Iron Man* films). Many are quick to praise him for taking these kinds of financial risks to develop valuable transport infrastructure, but of course the rewards for his company could be huge – and we'll end up with an electric charging network under private, not public, control. There's a wider point here, too: should the introduction of sustainable technology be reliant on the whims of billionaires? What's to stop Musk getting bored with electric vehicles and pouring all his resources into his space exploration company SpaceX instead?

This increasing reliance on companies, not governments, as providers of energy services and infrastructure is driven by a global economic system

A thought experiment

Some participants in a January 2015 event in London called 'Imagining Energy Democracy' tried to envision what a democratic energy system for Britain might look like:

- Some micro-grids and community schemes generating and storing their own off-grid power.
- A publicly owned national grid to which most households and businesses are connected, responsible for balancing supply and demand, using large amounts of energy storage.
- Solar, wind, wave, tidal stream and hydropower from a diverse mixture of local energy co-operatives, council-owned energy companies, community-owned projects and not-for-profit firms connected to the national energy grid.
- A large number of offshore wind turbines also providing energy to the grid, owned by a nationalized Offshore Wind Company. All surplus income from this company is recycled into supporting energy efficiency, sustainable transport, reliable grid connections and community renewable start-ups.
- Households pay their bills through their local energy co-operative, council-owned company or community energy scheme; or, failing that, a publicly owned energy distribution company acting as a default supplier, recycling its income back into energy schemes for the common good.

The event was organized by Fuel Poverty Action, Global Justice Now (formerly WDM) and Platform.

based on market 'liberalization', profit maximization and endless growth. It's a trend that we need to reverse if we want renewable energy truly to be a force for good.

Luckily, alternative models are appearing all over the world. Renewable-energy co-operatives have hundreds of thousands of members and are building and installing their own solar, wind and small-scale hydro projects from Indonesia to Costa Rica. They own three-quarters of Denmark's wind turbines, and are growing rapidly in Spain, Britain and elsewhere; in Germany, more than half of renewable electricity generation is owned by citizens, co-operatives and community groups.

The energy industry has been taken back into public hands by democratically elected governments in Venezuela and Bolivia. The popular state-owned energy system in Uruguay has had real success in expanding energy access, and is now working on efficiency and wind-power projects. Interest in locally controlled energy has been reignited in Europe by energy referendums in two major German cities (Hamburg and Berlin – see box above). However, it won't be possible to achieve true energy democracy without changes to our wider political and economic system. What's the use of campaigning for publicly owned energy if the national or local government is corrupt, undemocratic, or heavily influenced by vested interests?

We need to pursue the democratization both of our energy and our politics in parallel. In fact, we should see this as an opportunity, because these projects can support and mutually reinforce each other. Unaccountable corporate energy systems give powerful vested interests excessive influence over everything from household spending to government policy. Breaking the power of the fossil-fuel corporations and big utility firms, and creating new income streams for communities, co-operatives and the public sector, will open up all kinds of new spaces for democratic change.

To make this happen, we need to pick some

ambitious but achievable short-term goals that catalyze further change. There are lessons to be learned from Germany, Denmark and Bolivia, where government support for renewable-energy co-operatives has led to a genuine transfer of power towards the grassroots. Bringing energy industries back under national or local control could be a valuable step, if combined with other democratic reforms; Norway, Denmark and Uruguay all have strong representation from workers on their national energy bodies.

To achieve these goals, we will need to reduce the power and influence of the fossil-fuel companies, kicking their representatives out of government and moving subsidies away from polluting fuels and towards clean energy. We need our universities, pension funds and religious institutions to move their investment funds out of fossil fuels and into cleaner alternatives: not corporate renewable schemes but community energy, sustainable local transport and energy-efficiency projects.

We can't just sit back and expect the falling price of solar and wind to sweep away the old energy order. Renewable energy could be a powerful tool for dismantling the current failed system – but we need to use it wisely, and not let it fall into the wrong hands!

1 nin.tl/riskyenergy **2** International Energy Agency. **3** International Energy Agency. **4** nin.tl/guardianklein **5** See, for example, the *Guardian* nin.tl/takeoverpush **6** nin.tl/desertechgrab **7** nin.tl/hamburg-buy-back and nin.tl/berlinpeoplepower

12 Making it happen

> *'You cannot protect the environment unless you empower people.'*
> – Wangari Maathai

We don't need to burn fossil fuels. We can power the world without them.

This is lucky, because the fossil-fuelled future that we're currently heading for – as predicted by the International Energy Agency[1] – would be a pretty terrifying one. We'd almost certainly be locked into disastrous runaway climate change, and would experience ever more serious floods, storms, heatwaves, droughts, extinctions, collapsing food supplies and the loss of millions of people's homes, lives and livelihoods, within the lifetimes of most people alive today.

Whatever people might tell you, alternative futures are available. In this book, I've tried to give you a rough idea of how a future powered by renewable energy might shape up. In this future:

- We transition completely away from burning fossil fuels over the next 25 years, and thus massively reduce our CO_2 emissions. This gives us a good chance of avoiding runaway climate change.
- Global politics is no longer corrupted and distorted by the power of oil-producing companies and states, and millions of people no longer suffer from the negative effects of local fossil fuel extraction.
- We have a much more equitable world, where everyone has access to enough energy for a good quality of life.
- We don't need to use controversial technologies like nuclear power, carbon capture, geoengineering, giant new dams or corporate biofuels. Instead, we rely on cleaner, safer and more popular options such as wind and solar power.

This future is possible using existing technology. However, it requires some significant changes in our economy and society.

How likely is this second future? It requires a major global movement for change – but most people in the world claim to support this kind of shift in energy use. Meanwhile, the first future is based on some pretty extraordinary assumptions of its own – it assumes that the fossil-fuel companies really can extract all the oil from the tar sands, deepwater rigs and Arctic waters without causing any major catastrophes. It assumes that gas fracking, nuclear power, and biofuels will really generate as much energy as those industries claim, and that communities affected by all this destructive activity won't take a stand to stop them. It assumes that massive global energy inequality will continue unchallenged for the next 20 years, and that the entire human race will simply allow the world to slide into irreversible climate change. How likely does all of that sound?

We're standing at a crossroads. It's time for humanity to make a choice. Do we sit back and allow fossil-fuel companies and oil-friendly governments to dig, drill and frack us into a dark and dirty future? Or do we stand together with communities around the world to stop these extreme energy projects, and head down a different path into a safer, fairer energy future?

Which will *you* choose?

What you can do

The barriers to a clean-energy future are to do with politics, not technology. This is good news, because it means that we can all do something about it. Our economies and political systems were made by people, and can be changed by people.

What will that take? Well, if we look at the history of social change, the necessary conditions for major political shifts are:

- An active minority with right on their side, who are

willing to raise their voices and put themselves on the line to create change.

- This minority needs to be bigger, louder, or exert more leverage than the active minority at the other end of the scale, who benefit from the status quo and try to defend it.
- A majority who are willing to be dragged along with the 'good' minority. They don't need to be actively calling for the change, just willing to go along with it.

Sociologists and historians tell us that only rarely does a social movement actively involve more than five per cent of the population. For example, only five per cent of the US population played any active role in the civil-rights movement in the 1960s – and yet that movement created significant political change.

So we don't need everyone to become clean-energy activists. We just need enough people to take meaningful, strategic action, to exert greater political and social pressure than the obstructive minority, while building enough quiet support in the population at large to allow the change to happen.

So what does that mean?

Five per cent of the population is still a lot of people. The power of the fossil-fuel industry means we have a real fight on our hands. We need a lot more people to switch from passive acceptance of the problem to actively fighting to change the situation. We need a major movement for change, to take on the vested interests and win.

Here are some things you can do to help make this happen:

1. Reduce the power of the vested interests. All over the world, people are working to reduce the influence of the fossil-fuel industry over our politics and society. The fossil-fuel divestment movement, initiatives to kick dirty money out of politics, and campaigns to stop oil sponsorship of arts and culture could all benefit from your support. Campaigns to bring privatized energy back under public or community control are another

vital part of the picture.
2. Support the battles at the frontline. Communities directly affected by fossil-fuel extraction – and dirty 'renewables', like big dams, industrial biofuels and biomass – are fighting back against these destructive industries all over the world, and have already won some important victories. They're not just defending their own lands and livelihoods, but all of our shared futures, because every time they win they close off another swathe of fossil fuels or forests from exploitation. This adds to the costs of dirty energy and makes the cleaner alternatives more likely to happen instead.
3. Support energy reduction projects and campaigns. Better public transport and cycling, mass home insulation, local food initiatives, recycling schemes: we need all of these things to be seriously ramped up across the industrialized world. At the same time, we need campaigns to challenge our culture of overconsumption, and fight back against the expansion of roads and runways.
4. Make the links. The same companies and politicians that are pushing us into this dirty energy future are also responsible for a host of other injustices. If clean-energy campaigners can find common ground – and ways to actively support – all the people working for economic justice, equality and peace, then all of us are more likely to succeed.
5. Get some clean alternatives up and running. Even if you can't install your own renewable energy sources, perhaps you can work together with others to set up community energy projects, co-operatives and purchasing groups.

None of these things are quick or easy solutions, but if enough of us become involved then they'll be meaningful, effective and exciting. A better future really is possible, but it's not going to happen without our involvement. So what are you waiting for?

1 We present this fossil-fuelled future visually alongside the cleaner alternative at twoenergyfutures.org

Some useful weblinks for action

International
Fossil Free gofossilfree.org
Indigenous Environmental Network ienearth.org
International Rivers internationalrivers.org
La Via Campesina nin.tl/peasantfarmers
Trade Unions for Energy Democracy unionsforenergydemocracy.org

Australia
Australian Forests and Climate Alliance forestsandclimate.org.au
Embark embark.com.au
Greenpeace greenpeace.org/australia
Quit Coal quitcoal.org.au

Britain
Centre for Alternative Technology – how to set up a community renewable scheme nin.tl/communityrenewables
Global Justice Now nin.tl/climateenergycampaign
Transition Network transitionnetwork.org
We Own It weownit.org.uk
Fuel Poverty Action fuelpovertyaction.org.uk
Art Not Oil Coalition artnotoil.org.uk
BP or not BP bp-or-not-bp.org
Rising Tide risingtide.org.uk
Platform platformlondon.org
Community Reinvest communityreinvest.org.uk
Biofuel Watch biofuelwatch.org.uk

Canada
IdleNoMore idlenomore.ca
Greenpeace nin.tl/Canenergycampaign
Council of Canadians canadians.org/energy
Community Energy Partnerships Program nin.tl/cepresources

New Zealand/Aotearoa:
Coal Action Network Aotearoa nin.tl/coalaction
Co-operative Business NZ nin.tl/startingcoopNZ

United States
Center for Social Inclusion nin.tl/energydemocracyideas
Community Power Network nin.tl/communitypower
Energy Justice Network energyjustice.net/biomass
Oil Change International priceofoil.org and dirtyenergymoney.com
Transition United States transitionus.org

Picture credits (1.1 = Chapter 1, Figure 1)

1.1 Infrogmation of New Orleans/Creative Commons; 1.2 (top) Worklife Siemens/Creative Commons; 1.2 (left) Torresol Energy; 1.2 (right) Desertec UK; 1.3 © National Energy Foundation 2015; 1.4 Energy2014/ Creative Commons; 1.5 Avecendrell/Creative Commons; 1.6 beatriceco. com; 1.7 n/a; 2.1 Creative Commons; 2.2 Uberprutser/Creative Commons; 2.3 commons.wikimedia; 2.4 commons.wikimedia; 2.5 n/a; 2.6 commons.wikimedia; 2.7 Valentin Angerer/Altaeros Energies; 3.1 commons.wikimedia; 3.2 Tomia/Creative Commons; 3.3 commons. wikimedia; 3.4 image courtesy of SEPA. All rights reserved. Used with permission. 3.5 GrahamColm/Creative Commons; 3.6 LeGrand Portage/Rehman/Creative Commons; 3.7 n/a; 4.1 Lawrence Livermore National Laboratory; 4.2 Andrew Kokotka; 4.3 commons.wikimedia; 4.4 n/a; 4.5 commons.wikimedia; 5.1 TW Thorpe/ecowavepower. com; 5.2 Statkraft Development AS; 5.3 Swansea Lagoon; 6.1 n/a; 7.1 Vortexrealm/Creative Commons; 7.2 Matt Davis/Creative Commons.

Index

Page numbers in **bold** refer to main subjects of boxed text. Those in *italic* refer to illustration captions.